SYNTHETIC APERTURE RADAR FOR SURFACE MOTION ESTIMATION

TOPICS IN ADVANCED GEOINFORMATICS

Print ISSN: 3041-0770
Online ISSN: 3041-0789

Series Editors: Jianya Gong *(Wuhan University, China)*
Huayi Wu *(Wuhan University, China)*

This book series aims to comprehensively cover the theory, methodology and applications of geospatial information science (GIScience), which mainly includes advanced topics on geographic information system (GIS), remote sensing (RS), and global navigation satellite system (GNSS).

The books in this series will include monographs and graduate level textbooks aimed at young researchers and postgraduates specialising in remote sensing, surveying and mapping, geography, urban planning and environmental science, global climate change and related majors.

Published

Topics in
Advanced
Geoinformatics
Volume 3

SYNTHETIC APERTURE RADAR FOR SURFACE MOTION ESTIMATION

Timo Balz
Wuhan University, China

高等教育出版社
HIGHER EDUCATION PRESS

World Scientific

Published by

Higher Education Press Limited Company
4 Dewai Dajie, Xicheng District
Beijing 100120, P. R. China
and
World Scientific Publishing Co. Pte. Ltd.
5 Toh Tuck Link, Singapore 596224
USA office: 27 Warren Street, Suite 401-402, Hackensack, NJ 07601
UK office: 57 Shelton Street, Covent Garden, London WC2H 9HE

Library of Congress Control Number: 2025001953

British Library Cataloguing-in-Publication Data
A catalogue record for this book is available from the British Library.

Topics in Advanced Geoinformatics — Vol. 3
SYNTHETIC APERTURE RADAR FOR SURFACE MOTION ESTIMATION

ISBN 978-981-12-9853-0 (hardcover)
ISBN 978-981-12-9854-7 (ebook for institutions)
ISBN 978-981-12-9855-4 (ebook for individuals)

For any available supplementary material, please visit
https://www.worldscienti ic.com/worldscibooks/10.1142/13995#t=suppl

Desk Editors: Nambirajan Karuppiah/Steven Patt

Typeset by Stallion Press
Email: enquiries@stallionpress.com

About the Author

Timo Balz was born in Stuttgart, Germany. He received the Diploma degree in geography and the Doctoral degree in aerospace engineering and geodesy from the University of Stuttgart, in 2001 and 2007, respectively.

From fall 2001 to the end of 2007, he was a Research Assistant at the Institute for Photogrammetry, University of Stuttgart. Between 2004 and 2005, he was a Visiting Scholar at Wuhan University, Wuhan, China. From 2008 to 2010, he was a Postdoctoral Research Fellow at the State Key Laboratory of Information Engineering in Surveying, Mapping and Remote Sensing (LIESMARS), Wuhan University. From 2010 to 2015, he was an Associate Professor for Radar Remote Sensing with LIESMARS. Since 2015, he has been a Full Professor at LIESMARS. Since 2021, he has served as the Vice-Director of the International Academy of GeoInformation, Wuhan University.

He serves as an Associate Editor for the *IEEE Geoscience and Remote Sensing Magazine* and *Remote Sensing*. He is a member of the editorial board of *Geo-Spatial Information Science* and the *Journal of Digital Earth*. Since 2016, he has been the Chair of two ISPRS Commission I Working Group on SAR from 2016 to 2022 and again from 2022 to 2026. He has authored and co-authored more than 150 scientific articles in journals, books, and conference proceedings.

Timo Balz's research interests include surface motion estimation with SAR, data visualization, SAR geodesy, and the use of SAR data to support archaeological prospections.

Acronyms

APS	Atmospheric Phase Screen
ASI	Italian Space Agency
D-InSAR	Differential SAR Interferometry
DEM	Digital Elevation Model
DLR	Deutsches Zentrum für Luft- und Raumfahrt (German Aerospace Agency)
DSM	Digital Surface Model
ESA	European Space Agency
ESD	Enhanced Spectral Diversity
FFT	Fast Fourier Transform
FM	Frequency Modulation
GNSS	Global Navigation Satellite System
IFFT	Inverse Fast Fourier Transform
InSAR	Interferometric Synthetic Aperture Radar
LAMBDA	Least-squares AMBiguity Decorrelation Adjustment
LOS	Line-Of-Sight
PRF	Pulse Repetition Frequency
PS	Permanent Scatterer
PSC	Permanent Scatterer Candidate
PSI	Permanent Scatterer Interferometry
PTOT	Point-Target Offset Tracking
Radar	RAdio Detection And Ranging
RCS	Radar Cross Section
SAR	Synthetic Aperture Radar
SBAS	Short BAseline Subset
SCR	Signal-to-Clutter Ratio
SD	Spectral Diversity

SNR   Signal-to-Noise Ratio
SRTM   Shuttle Radar Topography Mission
StaMPS   Stanford Method of Persistent Scatterers
STUN   Spatio-Temporal Unwrapping Network
TOPS   Terrain Observation with Progressive Scanning

List of Symbols

A	Area
A_e	Effective antenna area
B	Bandwidth (range of the system's) bandwidth
$B_\perp$	Perpendicular baseline
c	Speed of light
D_A	Amplitude dispersion index
f_c	Radar center frequency
f_s	Range sampling frequency
G	Antenna gain
j	Imaginary unit
K_a	FM rate
K_r	Chirp rate
l_{ra}	Length of the real antenna
l_{sa}	Length of the synthetic antenna
P_n	Power of the noise
P_r	Received power
P_t	Transmitted power
r	Range (distance between sensor and object)
T_n	Noise temperature
v_{linear}	Linear LOS velocity of the target
v_{los}	LOS velocity of the target
v_{sensor}	Velocity of the sensor
λ	Wavelength
$\hat{\gamma}$	Estimation of the temporal coherence
δ_{az}	Spatial resolution in azimuth
δ_{grd}	Spatial resolution in ground-range
δ_{rg}	Spatial resolution in slant-range

δ_{sa}	Spatial resolution in azimuth of the synthetic antenna
θ_{inc}	Incidence angle
σ	RCS
σ^0	Percentage of the backscattered energy from an object
τ	Pulse length
τ_c	Chirp length
ϕ	Phase
ϕ_{atmo}	Phase contribution from the atmosphere
ϕ_{motion}	Phase contribution from motion
ϕ_{noise}	Phase contribution from noise
ϕ_{orbit}	Phase contribution from orbit estimation errors
ϕ_{res}	Phase residual
ϕ_{topo}	Topographic phase
φ	Wrapped phase
φ_{az}	Angular resolution in azimuth
φ_{sa}	Angular resolution in azimuth of the synthetic antenna

Contents

Part II   DEM Generation

Chapter 7 Alternative Approaches to PSI   **143**

Chapter 8 Distributed Scatterer Interferometry   **157**

Chapter 9 Pixel Tracking and Point-Target Offset Tracking   **165**

Chapter 10 Motion from SAR Geodesy   **173**

Part I   SAR Basics

Chapter 1

History of Radar Remote Sensing

Between 1886 and 1889, Heinrich Rudolf Hertz conducted a series of experiments at the Technische Hochschule in Karlsruhe, Germany, proving the existence of electromagnetic waves. He first let electric sparks jump across a narrow air gap between two charged metal spheres. For his experiment, he added a detector, a simple adjustable loop of wire, and a reflector, a large zinc sheet (Hertz, 1888). With this setup, he demonstrated that an electromagnetic wave is formed from the metallic spheres and reflected by the zinc sheet before it can be detected by his simple detector. This experiment proved the ideas that Michael Faraday developed in the 1830s and James Clerk Maxwell formalized in the 1860s.

Their achievements opened our minds to the concepts of the electromagnetic field, uniting electricity, magnetism, and light into one theory. The realization that light is just another form of electromagnetic wave is at the foundation of remote sensing as a science.

Letting the history start with Hertz is certainly arbitrary though. As it always is in science, the works of him, Faraday, and Maxwell, are all built on the work of others. Anastasia Volta and Luigi Galvani and their work in understanding and controlling electrical currents are certainly crucial here. Galvani discovered in 1780 that the muscles of dead frogs twitch when struck by an electric spark. Volta repeated Galvani's experiment, but believed that the effect was not intrinsic to the animal, but dependent on the metallic cable. Volta's work led to the development of an early battery that later on was used in many experiments.

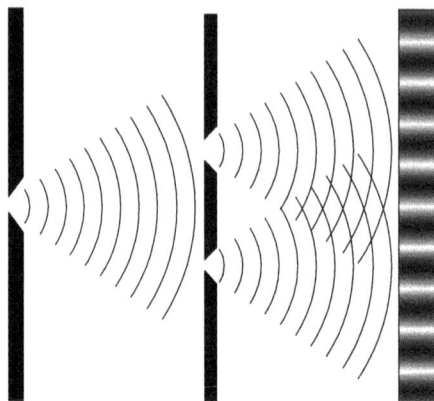

Figure 1.1   Young's double-slit experiment.

Thomas Young's double-slit experiment in 1801 would also be a good starting point. By directing light through two slits with a distance of a few centimeter between each other and illuminating a screen, Young demonstrated the wave nature of light (see Figure 1.1). Assuming light to consist of particles, as the previous theory of light did, the screen should have become smoothly lit; instead, it showed a pattern of bright and dark areas. These are caused by the interference of the electromagnetic waves from the two slits, thus proving the wave nature of light (Young, 1802).

Young also proposed that light, and other electromagnetic waves, are transverse waves, implying that the amplitude of the wave is perpendicular to the traveling direction of the wave. This is the polarization of an electromagnetic wave and is in contrast to longitudinal waves, like sound, where the amplitude is formed in the traveling direction of the wave.

Immanuel Kant (1786) presented a dynamical theory of matter and its forces, with attraction and repulsion forces. It gave the theoretical foundation for a unified treatment of all forces, ergo including electricity and magnetism. Von Schelling's (1797) work in the *Naturphilosophie* extended that with a unity of all forces, and the forces just representing a form of unified forces, which could be converted from and into each other, forming the first ideas of a unified field theory. Motivated by the *Naturphilosophie*, Oersted (1820) showed the direct influence of an electric current on a magnetic needle and the circular electromagnetic forces, although he did not yet combine them into one force.

This task was completed by Maxwell (1865) in what was later to be known as Maxwell's equations. The form in which they are nowadays typically shown was in 1885 though in the work of Oliver Heaviside, who reduced the original 20 equations to four using vector terminology.

$$\nabla \cdot \boldsymbol{E} = \frac{\rho}{\varepsilon_0} \tag{1.1}$$

$$\nabla \cdot \boldsymbol{B} = 0 \tag{1.2}$$

$$\nabla \times \boldsymbol{E} = -\frac{\partial \boldsymbol{B}}{\partial t} \tag{1.3}$$

$$\nabla \times \boldsymbol{B} = \mu_0 \left(\boldsymbol{J} + \varepsilon_0 \frac{\partial \boldsymbol{E}}{\partial t} \right) \tag{1.4}$$

This form uses differential equations with the nabla symbol ∇, denoting a three-dimensional gradient operator that can also be reformulated using integral equations over volumes and surfaces. $\boldsymbol{E}$ is a vector describing the electric field, $\boldsymbol{B}$ is a vector describing the magnetic field. ε_0 is the permittivity of free space and μ_0 is the permeability of free space. ρ is the total electric charge density and $\boldsymbol{J}$ is the total electric current density.

These four equations are also known as Gauss's law, Gauss's law for magnetism, Faraday's law, and Ampere's law with Maxwell's addition. These equations are fundamental for radar remote sensing. They describe electric and magnetic fields, and their charges, currents and changes. A consequence of them is the theoretical foundation of the speed of light c as well as an understanding that all electromagnetic waves, including microwaves, travel at the speed of light:

$$c = \frac{1}{\sqrt{\varepsilon_0 \mu_0}} \tag{1.5}$$

With Maxwell's equations formulating the theory and Hertz's experimental proof, the application of this was only demonstrated later on, for example, by Guglielmo Marconi, who successfully transmitted radio waves over the Atlantic in 1902.

The idea of echolocation, which is the method of transmitting waves and positioning objects by the returned energy, has also already been known around 1900. It was initially developed using sound waves, e.g. from a horn, listening to the returning echo and measuring the time difference to establish the direction and distance of, e.g., icebergs. Nikola Tesla showed the possibility of using radio waves for this as well, but never worked out the details. The first working solution was patented by Hülsmeyer (1904). Marconi presented a more advanced approach using detection and ranging with radio waves in 1922. The first pulsed radar system was developed by Robert Alexander Watson-Watt, who received the patent for his radio detection and ranging (Radar) device in 1935.

The first Radar systems used very long wavelengths and therefore required huge antennas. To make them less bulky and usable on ships and airplanes, it was necessary to reduce their size. This became possible with the cavity magnetron, which is a high-powered vacuum tube generating microwaves using the electro-magnetic interactions moving past open metal cavities functioning as a resonating body. Various teams around the world were working on this or similar devices. In Germany and other Axis countries, the Klystron was the preferred device for microwave generation due to its better frequency stability. That was based on the work of Arsenjewa-Heil and Oskar Heil, although simultaneously and, most probably, unaware of each other, Russel Varian and Sigurd Varian developed and published a prototype at Stanford University (Varian & Varian, 1939).

The cavity magnetron developed by John Randall and Harry Boot from the University of Birmingham was the breakthrough though. Their device allowed multi-kilowatt pulses of 10 cm wavelength, allowing for smaller antennas, reducing the size of Radar systems. Based on this, the first airborne mapping Radar was the H2S Radar used by the Royal Air Force for navigation and night-time bombing for the first time in 1943.

The next breakthrough for Radar as remote sensing device was the development of the Doppler beam sharpening by Carl Wiley (1951) while he was working for Goodyear Aircraft. His work allowed the increase of the spatial resolution of the systems. For this work, he is credited as the inventor of Synthetic Aperture Radar (SAR).

The first large mapping project using airborne SAR was then carried out in Panama in 1967. The success led to further campaigns, for example, in Venezuela. The cloud-penetrating capability of microwaves makes SAR especially useful for mapping tasks in tropical areas with almost constant cloud coverage, as it allows for higher altitudes and faster mapping.

The first civilian spaceborne radar imaging system was Seasat launched in 1978. The system failed prematurely after only 110 days, but nevertheless demonstrated the immense capability of spaceborne Radar systems. Seasat was followed by Shuttle Imaging Radar (SIR) missions in 1981 (SIR-A), 1984 (SIR-B), and 1994 (SIR-C/X-SAR).

In 1991, the European Remote Sensing Satellite-1 (ERS-1) was launched. ERS-1 pioneered SAR remote sensing due to its capability of SAR interferometry and differential SAR interferometry. With the launch of ERS-2 in 1994 and thanks to the extremely long life of ERS-1, which stayed in operation until 2000, ERS-1 and ERS-2 allowed SAR interferometry with short temporal baselines. The success of the ERS mission led to an ongoing investment of the European Space Agency (ESA) in C-band systems, with ENVISAT's ASAR, launched in 2002 and staying operational until 2012.

The success of the ERS mission in DEM generation from SAR interferometry led to the development of the Shuttle Radar Topography Mission (SRTM). SRTM was a joint mission of National Aeronautics and Space Administration (NASA), the Deutsches Zentrum für Luft- und Raumfahrt (DLR), and the Italian Space Agency (ASI). The SRTM mission was the first fixed baseline single-pass spaceborne interferometric synthetic aperture radar (InSAR) system. Its goal, the creation of a global digital elevation model, was achieved in 10 days with an unprecedented accuracy and coverage.

In June 2007, the DLR launched their first high-resolution SAR satellite: TerraSAR-X, which was followed in 2010 by TanDEM-X, an identical satellite. Both fly in a close formation, apart only by a few hundred meters, allowing for bi-static SAR acquisition, ideal for DEM generation. This mission produced a high-resolution vertical surface model of the entire land surface of the world in 12 m resolution and with a vertical accuracy < 2 m.

In this short overview over the development of spaceborne SAR systems, several milestone satellites have not yet been included. There is ASI's COSMO-SkyMed constellation, which similar to TerraSAR-X launched in 2007 and also provides high-resolution SAR images in the X-band.

The Japan Aerospace Exploration Agency (JAXA) launched their first SAR satellite, JERS-1, already in 1992, shortly after the launch of ERS-1. JAXA continues its commitment to L-band SAR systems with ALOS PALSAR (2006–2011) and recently with the PALSAR-2 system launched in 2014.

The Canadian Space Agency (CSA) launched the first commercial SAR satellite, Radarsat-1, in 1995. Radarsat-1 outlived its life expectancy and stayed in service for more than 17 years until 2013. Radarsat-2 was launched in 2007 and is still operational. In 2019, the Radarsat Constellation Mission (RCM) consisting of three satellites was launched successfully.

ESA launched the Sentinel-1 mission in 2014 with Sentinel-1A followed by Sentinel-1B in 2016. The Sentinels are remarkable SAR satellites that provide global coverage with a 12-day repeat cycle. Thanks to the open-data policy of the Copernicus mission, of which Sentinel-1 is a part, the global data are freely available, offering a globally available treasure of SAR data time series.

Recently, new agencies launched their own SAR satellites, such as the HJ-1C from China launched in 2012. With GaoFen-3 (GF-3), China launched another SAR satellite in 2016 and has several more SAR systems in preparation.

Republic of Korea joined the club of SAR satellite owners in 2016 with KOMPSAT-5, as did Spain and Argentina in 2018 with PAZ and SAOCOM.

Besides the traditional national space agencies, private companies and start-ups plan to launch or have already launched SAR satellites, such as ICEEYE run by a Finish-Polish company, or Capella Space from the USA. They, and also several of the traditional space agencies, plan on increasing the SAR constellations, with several similar SAR satellites in a constellation, increasing the revisit frequency. Therefore, the increase in available SAR satellites shown in Figure 1.2 is only just the beginning and a further increase can be expected in the future.

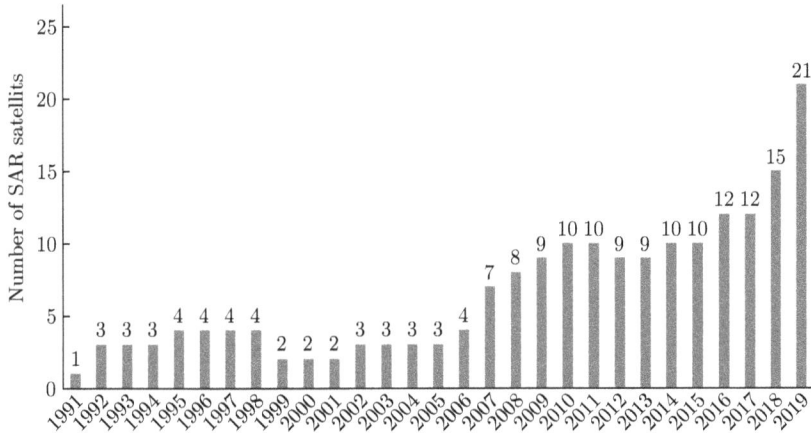

Figure 1.2   Civilian SAR satellites in orbit.

Chapter 2

Introduction to Synthetic Aperture Radar

Radar and especially the Synthetic Aperture Radar (SAR) are important remote sensing sensors. Before learning to use SAR for surface motion estimation, the basics of radar and SAR must be understood. An example of ship detection is used as an application to base this introduction on a practical application.

The detection of ships is a fundamental remote sensing task with many applications in maritime management and security-related operations. Remote sensing systems for ship detection should discern between water and ships to ensure a high detection rate. Furthermore, ships should be detectable under all weather conditions, at day and night, to allow for a continuous surveillance. Radar systems are ideal for this and, in fact, we can say, radar systems are built for this application. The very first radar system, developed by Hülsmeyer (1904), was specifically designed to detect metallic objects on water under bad weather conditions.

2.1 Technical Considerations for a Ship Detection Remote Sensing System

2.1.1 Electromagnetic Spectrum

A ship detection system should be able to detect ships under all weather conditions at day and night. This will not be possible using a system working in the visible range of the electromagnetic spectrum because

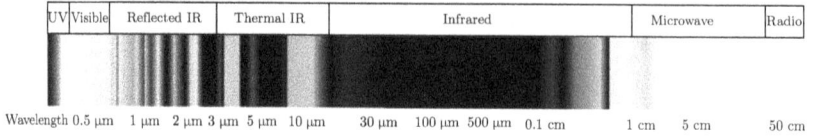

| UV | Visible | Reflected IR | Thermal IR | Infrared | Microwave | Radio |

Wavelength 0.5 μm   1 μm   2 μm 3 μm 5 μm   10 μm   30 μm   100 μm 500 μm   0.1 cm   1 cm   5 cm   50 cm

Figure 2.1   Electromagnetic spectrum: White represents full atmospheric transmission, whereas black represents no transmission.

these short wavelengths cannot penetrate through clouds and fog. Furthermore, passive systems depending on the Sun as a light source will not be feasible because such systems will only work during the day. Instead, we need an active system which operates in a frequency range that allows transmission through clouds and fog.

As shown in Figure 2.1, the visible range has good transmission capability through the atmosphere. In the longer optical wavelengths, such as reflected infrared and thermal infrared, certain atmospheric windows exist, where wavelengths can be transmitted; however, it is only at the microwave spectrum, i.e. from 1 cm to about 1 m wavelength, where we again see very high transmissibility. In fact, electromagnetic waves in the microwave spectrum can penetrate clouds and fog, and therefore can allow radar systems to 'see' through clouds.

A radar system is an active system. The electromagnetic waves are transmitted by the radar system and received by the same or another system. It is therefore independent from the sunlight and can operate during the day and night. It needs, however, a larger amount of energy to operate, unlike passive remote sensing systems. A spaceborne SAR system therefore depends on the Sun to provide energy to operate. These properties of active microwave systems allow us to use them under (almost) all weather conditions, at day and night, fulfilling the basic requirements for ship detection applications.

2.1.2   Microwave Scattering Behavior

Having a system that can operate under different conditions is great, but another important requirement is the ability of the system to discriminate between our objects of interest, i.e. ships, and the background, i.e. water.

Microwaves for Ship Detection

A radar system transmits electromagnetic waves in the microwave spectrum. We can differentiate between two classes of radar systems: mono-static systems and bi- or multi-static systems. In a mono-static system, the same antenna is used to transmit the electromagnetic signal and receive the backscattering. In a bi- or multi-static system, the signal is transmitted from one antenna and received by another antenna or multiple other antennae. If not specifically mentioned, we will consider only mono-static systems by default.

We can further differentiate between continuous wave radar systems and pulsed radar systems. In a continuous wave radar system, one antenna continuously transmits a radar signal, while the other antenna is continuously receiving signals. In a pulsed system, a radar signal is transmitted as a pulse and then the antenna is switched to receiving mode to collect backscattering before transmitting again. In the following discussion, we will only consider pulsed systems, although we could also design a ship detection system with continuous wave radar.

Given a pulsed mono-static radar system, a radar signal is transmitted and the signal is scattered back to the system. The strength of the backscattered signal, which is the signal transmitted from the radar, scattered at an object and received by the radar antenna, strongly depends on the physical appearance of the observed object. The roughness of the object surface as well as the orientation of the object surfaces toward the sensor are dominant factors.

In Figure 2.2, we assume reflecting surfaces with a high enough di-electrical constant, allowing for backscattering of the radar signal. Water, for example, has a very high di-electrical constant and strongly reflects microwave signals. However, considering a smooth surface, the radar signal is reflected away from the sensor and no signal is received in a mono-static configuration. This is the case for calm water surfaces. Having a rougher surface, parts of the signal are reflected

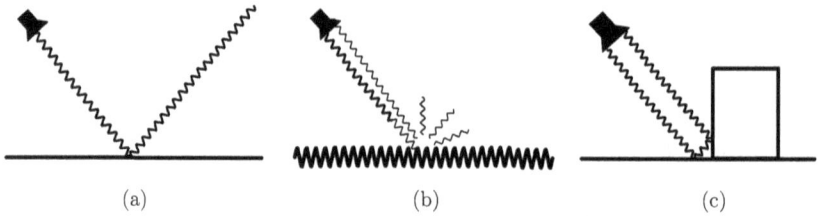

Figure 2.2   SAR reflection from (a) smooth surfaces, (b) rough surfaces and (c) double-bounce scattering.

back to the sensor, as in the case of turbulent water surfaces and vegetation covered areas in radar remote sensing images. However, having a smooth surface that reflects microwaves away can also lead to a very strong backscattering if these reflected microwaves are reflected again back to the sensor, for example, by a metallic object like a ship. This is the main reason for the strong discernibility of ships on water in radar remote sensing images. The radar waves are reflected away from calm water surfaces, leading to very low amplitude in the radar images from water, whereas the radar signal hitting a ship is reflected back to the sensor. Additionally, a signal that hits water in front of the ship can be reflected toward the ship and then back to the sensor. This so-called double-bounce effect can lead to very strong backscattering, causing ships to appear with a very high amplitude in radar images. This makes them easily recognizable against the low amplitude background of calm waters.

2.2   The Relation between Ship Size and Radar Backscattering

Instead of only detecting metallic objects in the water, an estimation of the object's size would be useful. The strength of the received radar signal can give us an indication on the size of the object, as larger objects can reflect more energy back to the sensor.

2.2.1 Radar Equation

Having an active sensor, like a radar system, has certain advantages. One of them is the ability to transmit a well-defined signal. Knowing the properties of the transmitted signal allows us to better understand the physical properties of the illuminated objects. This is expressed by the so-called radar equation, which describes the relationship between the transmitted and received power of a radar system. In the following, we will discuss the radar equation step by step. First, we have the transmitted power P_t given in Watt (W). At a given distance r, the power density is described as in

$$\frac{P_t}{4\pi r^2} \tag{2.1}$$

and given in $W \cdot m^{-2}$. This can be understood as the power times the inverse of the surface area of a sphere with a radius r. The so-called antenna directivity gain G is a dimensionless value calculated from the effective antenna area A_e and the system's wavelength λ with

$$G = \frac{4\pi A_e}{\lambda^2} \tag{2.2}$$

A_e is maximally the size of the antenna, but typically smaller. Thus, we can derive the power density at the scattering object with

$$\frac{P_t}{4\pi r^2} G \tag{2.3}$$

Having an object intercepting the signal with an area A, we can derive the power at the object in W with

$$\frac{P_t}{4\pi r^2} GA \tag{2.4}$$

An important value describing the backscattering properties of the object is σ^0 describing the fraction of the energy backscattered to the radar system. This signal travels back to the sensor, further reduced in its power by

$$\frac{1}{4\pi r^2} \tag{2.5}$$

We can understand this as the backscattering object acting as another antenna, with the power loss again being the inverse of the area of a sphere with radius r. The power density at the receiving antenna is

$$\frac{P_t}{4\pi r^2} \cdot G \cdot A \cdot \sigma^0 \cdot \frac{1}{4\pi r^2} \tag{2.6}$$

The power received at the antenna with an effective antenna area A_e is then described by

$$\frac{P_t}{4\pi r^2} \cdot G \cdot A \cdot \sigma^0 \cdot \frac{A_e}{4\pi r^2} \tag{2.7}$$

which gives us the final radar equation of the received power P_r:

2.1. Radar equation:

$$P_r = \frac{A_e G P_t}{(4\pi)^2 r^4} \sigma^0 A \tag{2.8}$$

Analyzing the radar equation, we can find that the received power P_r is strongly reduced in range with r^4. Furthermore, only two variables describe the object properties given by the area visible from the radar A and the fraction of the energy backscattered to the radar σ^0. Assuming similar material properties and therefore a similar fraction of the received energy being backscattered, the received power directly depends on the size of the object A and it is therefore possible to estimate the size of the illuminated object.

However, the illuminated size of an object is not identical to the size of the object. If we take the example of a ship again, A would strongly depend on the heading direction of the ship and the looking angle of the radar.

In most applications, we refer to σ^0 and A together as the radar cross section (RCS) σ of an object given in m^2. Since there can be huge differences in the backscattered power of radar images, logarithmic scaling is often preferred. The RCS is therefore often given in decibels (dB) with $dB = 10 \log_{10}(P)$, where P is the power ratio. In the case of the RCS, the power ratio is often formed by dividing the RCS by $1\ m^2$, so that the RCS in dB has no unit.

2.2.2 Controlling the Radar Backscattering of Ships

Radar is used in many maritime applications, and therefore, there is a certain interest in controlling the RCS of ships. Small ships have an interest in increasing their RCS to be better seen and to travel more safely. Military vessels may have an interest in reducing the RCS to improve their stealth capabilities. In both cases, controlling the RCS is very important.

Smaller vessels can increase their RCS using artificial radar targets. A Luneburg Lens (Gutman, 1954; Luneburg, 1944) allows for an increased RCS in all directions, whereas other targets show a more or less strong dependence on the looking direction. An advantage of artificial targets is that their theoretical peak backscattering can be calculated, which allows them to be used for radar system calibration.

For the calculation of the RCS of the artificial targets shown in Figure 2.3, we can use

$$RCS = \frac{4\pi}{\lambda^2} A_s^{\ 2} \tag{2.9}$$

where A_s depends on the size of the target and can be calculated.

(a)   (b)   (c)

Figure 2.3   Artificial targets: (a) mirror, (b) dihedral corner, and (c) corner cube.

2.2. Mirror:

$$A_s = a \cdot b$$

with a and b being the size of the rectangle forming the mirror.

2.3. Dihedral:

$$A_s = \sqrt{2} \cdot a \cdot b$$

with a and b being the sizes of the two rectangle forming the dihedral.

2.4. Corner Cube:

$$A_s = \sqrt{3} \cdot a^2$$

with a being the size of the corner cube sides.

On the other hand, to reduce the RCS, one would want to reduce the area of the ship reflecting back to the sensor, trying to reflect energy away from the sensor and use material that only reflects a very small fraction of the incoming radar signal. For a more detailed discussion on the radar cross section, readers may refer to the book by Knott *et al.* (2004).

2.2.3 Example for the Required Specifications to Detect Ships with a Radar Remote Sensing System

Example 2.1 We assume a spaceborne SAR system. Our system has a peak transmission power P_t of 5000 W. The sensor is 700 km away from the target and illuminates an area of 60 km^2. We assume an antenna directivity gain G of 40 000 and an antenna area of 10 m^2. For the fraction of the backscattered power σ^0, we assume 5%. According to the radar equation, the received power P_r can be calculated with

$$P_r = \frac{A_e G P_t}{(4\pi)^2 r^4} \sigma^0 A$$

Giving a $P_r = 1.58 \times 10^{-10}$ W. As we can see there is a huge difference between the transmitted power P_t and the received power P_r. □

As shown in the example, the echo is 135 dB below the transmitted power level. Therefore, only a tiny fraction of the transmitted power is received by a spaceborne radar system.

This has consequences for the system design. With such huge differences, a separation between the transmitted energy and the received energy is necessary, as both use the same antenna. Transmission power leakage into the receiving system has to be avoided, which is especially difficult for bi-static systems.

Considering ship detection, a high signal-to-clutter ratio (SCR) has to guarantee clear detection, i.e. the signal from the ship has to be much stronger than the signal of the background. To ensure a high RCS, a corner cube can be placed on the ship in our simplified example. For good discernibility, an SCR of 20 dB is desired.

For our calculation, we assume a C-band sensor with 5.15 cm wavelength and a pixel size of 10 m × 10 m. In the background, we assume very quiet sea where most of the signal is reflected away from the sensor, so that we assume −15 dB noise. We can calculate the background RCS with the following.

Example 2.2

$$\text{RCS}_{\text{background}} = 10 \text{ m} \times 10 \text{ m} \times 10^{-1.5} = 3.16 \text{ m}^2$$

or 5 dB.   □

For an SCR of 20 dB, our artificial target should therefore reach 25 dB peak signal strength, which is equivalent to an RCS of 316 m². The RCS of a target can be calculated with

$$\text{RCS} = \frac{4\pi}{\lambda^2} A_s^{\ 2}$$

so we need A_s of a certain size that we can calculate with.

Example 2.3

$$A_s = \sqrt{\frac{\text{RCS} \cdot \lambda^2}{4\pi}} \approx 0.26 \text{ m}^2$$

The corner cube is

$$A_s = \sqrt{3} \cdot a^2$$

so we need a corner cube with a side length a of

$$a = \sqrt{\frac{A_s}{\sqrt{3}}} \approx 0.39 \text{ m}$$ □

Therefore, a corner cube of 39 cm side length would, if illuminated from an appropriate direction, give the necessary 20 dB SCR to be clearly detected against a calm sea with a very weak backscattering. However, the sea is not always calm and waves can cause significant backscattering in SAR images. As shown in Figure 2.4, strong waves can lead to specular

Figure 2.4 Specular reflection from high waves.

Figure 2.5 TerraSAR-X image of Sansha acquired on 2019-01-26, showing backscattering from waves (© DLR, 2019).

reflection back to the sensor. Such wave formations show very high amplitude, as shown in Figure 2.5.

To guarantee an SCR of 20 dB even under such conditions, we should assume a higher background noise.

Example 2.4 If we assume, e.g. 1 dB backscattering from waves,

$$\text{RCS}_{\text{background}} = 10 \text{ m} \times 10 \text{ m} \times 10^{0.1} \approx 126 \text{ m}^2$$

the background already reaches 21 dB and we would need 41 dB backscattering strength from the corner cube. For the necessary RCS of 12 600 m², we would need a corner cube with

$$A_s = \sqrt{\frac{12\,600 \text{ m}^2 \times (0.0515 \text{ m})^2}{4\pi}} \approx 1.63 \text{ m}^2$$

$a \approx 0.97$ m side length. $\qquad\qquad\qquad\qquad\qquad\qquad \square$

All of these calculations do not include backscattering from the ship itself. Moreover, a corner cube does not provide strong backscattering from all looking directions, but, for example, an octahedron can be used to achieve high backscattering from almost all directions, which is widely used to increase the RCS of smaller ships.

As we can see from our example, it is quite easy to achieve a high SCR for ships in calm water and our system would be able to detect ships with corners of just 30 cm side length, which many ships will form without adding artificial targets. Even the 1 m corner size necessary to detect ships with high SCR in water with waves is commonly found on many ships, especially larger metallic vessels.

2.3   Distinguishing Different Ships with a Radar Remote Sensing System

Being able to measure the backscattering from a ship and differentiate it from the water is important. However, the spatial or angular resolution of a system is also very important, as it allows us to differentiate the backscattering and determine the number of ships and the type of ships in a given area. This is the spatial resolution of the radar system; there are, however, fundamental differences between the resolution of a radar system and the resolution of an optical system.

2.3.1 Radar Resolution

A radar system has two independent resolutions. First, there is the resolution in range, that is, the direction the signal is traveling. The radar system is transmitting a signal and the time from transmitting the signal to receiving the signal is used to determine the distance. Second, there is the resolution in the other direction, in remote sensing typically referred to as azimuth resolution. This depends on the angular resolution of the system.

As shown in Figure 2.6(a), the resolution of a real radar in azimuth depends on the angular resolution. The three black objects farther away from the sensor cannot be distinguished by the system, whereas closer to the sensor, in near range, we can distinguish the black point from the two gray points. In range, as shown on the right side of Figure 2.6, the returns of different objects in near and far range have different running times. The resolution in range depends on the possibility to distinguish between these pulses.

2.3.1.1 *Angular resolution of a radar system*

The angular resolution depends on the diffraction limit that can be derived from the Rayleigh criterion (Rayleigh, 1879) and can be approximated with

$$\alpha \approx \frac{\lambda}{A_d} \tag{2.10}$$

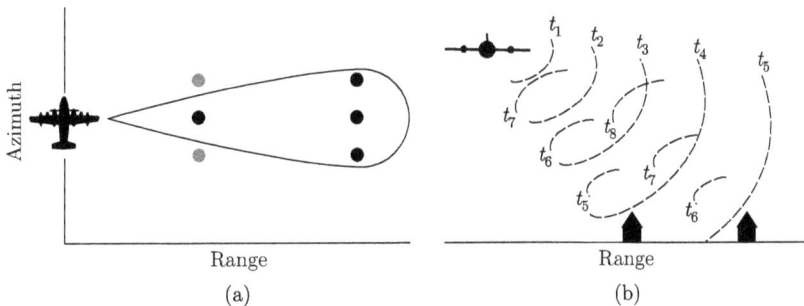

Figure 2.6 Resolution in azimuth (a) and range (b).

For the human eye working with $\lambda = 500$ nm and a diameter of approximately 5 mm, we can estimate the resolution in 700 km distance r with simple trigonometry, so that

$$\text{resolution} = \tan \alpha \cdot r \approx 70 \text{ m}$$

However, assuming now our spaceborne ship detection radar system working in C-band with $\lambda = 5.15$ cm and an antenna diameter of 10 m, we get a resolution of 3.6 km. Such a resolution would not allow us to distinguish different ships.

2.3.1.2   Range resolution of a radar system

In range direction, the resolution of a radar system depends on the time duration of a pulse τ. The resolution in slant-range δ_{rg}, i.e. the looking direction of the system, can be calculated with

$$\delta_{rg} = \frac{c\tau}{2} \tag{2.11}$$

where c is the speed of light. The factor of 2 is due to the traveling of the signal from the sensor to the object and back. In most applications, we are not interested in the slant-range resolution, but rather in the resolution of the system with respect to the ground, the so-called ground-range resolution. This resolution depends on the incidence angle θ_{inc}.

$$\delta_{grd} = \frac{\delta_{rg}}{\sin \theta_{inc}} = \frac{c\tau}{2\theta_{inc}} \tag{2.12}$$

Example 2.5   If we, for example, assume a pulse length $\tau = 40$ μs, we would see a slant-range resolution of

$$\delta_{rg} = \frac{299\,792\,458 \text{ m} \cdot \text{s}^{-1} \times 40 \times 10^{-6} \text{ s}}{2} \approx 5996 \text{ m} \qquad \square$$

Again, at a resolution of 6 km in range, we would not be able to distinguish ships. It is therefore necessary to increase the resolution in range. To increase the resolution in range, the pulse length needs to be reduced. There is a limit on the reduction of the pulse length though because it is very difficult to transmit a strong pulse in a very short time.

2.3.1.3 Using frequency modulated signals to improve range resolution

The range resolution can be improved by reducing the pulse length, which is limited by the technical difficulty when transmitting a strong pulse in a very short time. Alternatively, a longer pulse with a linearly increasing frequency over the pulse interval, also known as chirp signal, can be transmitted. The received signal is filtered by a match filter, fitting the reference signal, allowing us to separate the returns of the signal received with different frequencies, so that the signal can be separated by frequency. This allows a drastic increase in range resolution, but a larger frequency range is needed.

The effective τ of a chirp signal is

$$\tau = \frac{1}{B} \tag{2.13}$$

where B is the bandwidth of the signal around the center frequency:

2.5. SAR range resolution:

$$\delta_{rg} = \frac{c}{2B} \tag{2.14}$$

Example 2.6 So, for a bandwidth of 20 MHz,

$$\tau = \frac{1}{20 \times 10^6 \text{ Hz}} = 50 \text{ ns}$$

which corresponds to a slant-range resolution of

$$\delta_{rg} = \frac{299\,792\,458 \text{ m} \cdot \text{s}^{-1} \times 50 \times 10^{-9} \text{ s}}{2} \approx 7.5 \text{ m} \qquad \square$$

Using chirp signals, the resolution in range can be drastically increased to resolutions that easily allow users to separate ships. The limiting factor is the available bandwidth. The bandwidth in the radio spectrum is limited. Cell phone signals, TV signals, radio signals, etc. compete for bandwidth. The frequency range is therefore limited and band-dependent.

2.3.2   Side-Looking Airborne Radar Systems

Example 2.7   We use an airborne side-looking radar system. To allow for a better angular resolution, we use an X-band system with 3 cm wavelength. To allow for a good discernibility of larger ships, we want to have at least a 20 m ground resolution in the range and the azimuth direction. Our airborne system will fly at an altitude of 5000 m with an incidence angle of 45°.

Our sensor will use a chirp signal to improve the range resolution.

$$\delta_{rg} = \delta_{grd} \sin \theta_{inc} = 20 \sin(45°) = 14.1 \text{ m}$$

This requires τ of

$$\tau = \frac{2\delta_{rg}}{c} = 94 \text{ μs}$$

A bandwidth B of

$$B = \frac{1}{\tau} = 10.6 \text{ MHz}$$

With a wavelength λ of 3 cm, we have a radar center frequency f_c of

$$f_c = \frac{c}{\lambda} = 9.993 \text{ GHz}$$

If we use a bandwidth of 12 MHz, our frequency modulated radar signal will transmit from 9.987 GHz to 9.999 GHz.   □

The angular resolution of the system in azimuth φ_{az} can be estimated with

$$\varphi_{az} = \frac{\lambda}{l_{ra}}$$

where l_{ra} is the length of the real antenna. The azimuth resolution δ_{az} depends on the slant range r

$$\delta_{az} = r\frac{\lambda}{l_{ra}} \qquad (2.15)$$

Example 2.8 With a flight height of 5000 m and an incidence angle of 45°,

$$r = \frac{5000 \text{ m}}{\sin(45°)} = 3535 \text{ m}$$

We can derive the required antenna size from

$$l_{ra} = \frac{r\lambda}{\delta_{az}} = \frac{3535 \text{ m} \times 0.03 \text{ m}}{20 \text{ m}} = 5.3 \text{ m}$$

Attaching a 5.3 m long antenna to an airborne system is possible, but it would be a rather large antenna size. However, looking at the equations above, we can find a strong dependence of the antenna size to the range. Therefore, a spaceborne system would need a much larger antenna, for example, assuming a satellite in 700 km distance, the required antenna size l_{ra} increases to

$$l_{ra} = \frac{700\,000 \text{ m} \times 0.03 \text{ m}}{20 \text{ m}} = 1050 \text{ m} \qquad \square$$

An antenna size of 1 km is impossible to deploy in space. It is therefore necessary to further reduce the antenna size of the system. For this purpose, we can build an artificial antenna using the motion of the platform, a so-called synthetic aperture.

2.3.3 Synthetic Aperture Radar Systems

The main idea behind the synthetic aperture radar system is to increase the antenna size by forming a synthetic antenna along the trajectory of the sensor. That is to say, in order for a synthetic aperture radar (SAR) system to work, the sensor has to move during the image acquisition (Figure 2.7).

> **R** It is also possible for the sensor to remain stable at a position with the backscattering object moving in order to form a synthetic aperture. This is called inverse SAR (ISAR).

If we assume our sensor to move on a straight line, the SAR sensor is transmitting and receiving radar pulses along its trajectory with a certain pulse repetition frequency (PRF). By combining these pulses, it is possible to form a larger synthetic antenna from the pulses along the trajectory. The motion of the sensor is used to form a larger synthetic antenna out of the pulses of the smaller real antenna along the trajectory.

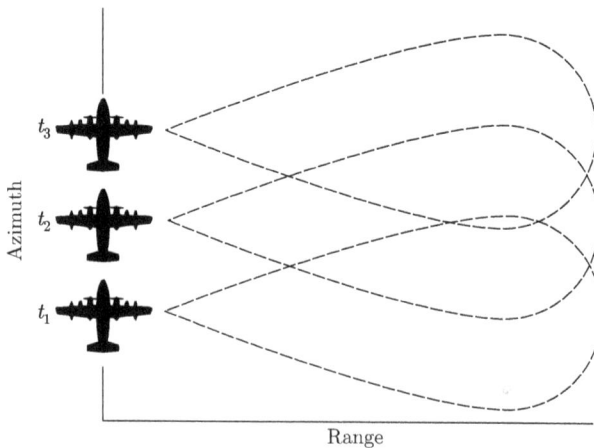

Figure 2.7 Forming of the synthetic antenna.

The length of this synthetic antenna l_{sa} depends on the angular resolution of the real antenna φ_{az} and the distance between the sensor and the object r. φ_{az} can be calculated from the equations on the azimuth resolution of the real antenna

$$l_{sa} = \varphi_{az} \cdot r = \frac{\lambda}{l_{ra}} r \qquad (2.16)$$

where l_{ra} is the length of the real antenna. For a synthetic antenna, the radar lobe φ_{sa} is just half the size due to the impulse compression, and therefore,

$$\varphi_{sa} = \frac{\lambda}{2l_{sa}} \qquad (2.17)$$

If we combine these equations, we can derive the resolution of the synthetic antenna δ_{sa}:

$$\delta_{sa} = r\frac{\lambda}{2l_{sa}} = \frac{\lambda l_{ra} r}{2\lambda r} = \frac{l_{ra}}{2} \qquad (2.18)$$

2.6. SAR azimuth resolution:

$$\delta_{sa} = \frac{l_{ra}}{2} \qquad (2.19)$$

This shows an astonishing effect: The resolution of the SAR system is independent of the distance between the sensor and the object and can be reduced to half of the length of the real antenna. That is to say, an SAR system can reach high resolutions in azimuth and range direction while the resolution is, in both directions, independent of the distance between the sensor and the object. Therefore, a spaceborne SAR system can reach the same resolution as an airborne SAR system.

Another way of understanding SAR systems considers the increase in azimuth resolution via the Doppler effect. By analyzing the frequency shift of the backscattering, we can determine the relative motion of the sensor with respect to the backscattering object. Therefore, we can determine if the sensor is moving toward an object or away from an object. Following this Doppler shift along the trajectory, the time of the sensor going toward to moving away from the sensor can be established, and therefore, the position of the object in azimuth can be determined.

SAR image focusing

SAR image acquisition is a two-step process. First comes the data acquisition, receiving the backscattering from the objects. From this received raw data, an image is created through the SAR focusing process. As there are many pulses recorded along the path, the contribution of a single point may be dispersed over 10^4–10^7 samples in the so-called SAR raw data image. This raw data has to be processed, the so-called SAR focusing, to get a meaningful SAR image.

The SAR focusing process is separated in azimuth and range direction. In range, the chirp signal has to be focused, while in azimuth, the synthetic aperture has to be synthesized and focused.

The first step is focusing in range. For this, a reference chirp in range is formed.

$$v(\tau) = \exp(j \cdot \pi \cdot K_r \cdot \tau^2), \quad -\frac{\tau_c}{2} < \tau < \frac{\tau_c}{2}$$

where j is the imaginary unit, K_r is the chirp rate and τ_c is the chirp length. $v(\tau)$ is formed with a step width $\Delta\tau$ of

$$\Delta\tau = \frac{1}{f_s}$$

where f_s is the range sampling frequency.

In pseudo-code, the focusing in range can be described as

2.7. Range focusing:
> **input** : Unfocused raw data
> **output:** Range compressed image
> **for** *line* ← 0 **to** *size$_{image}$* **do**
> > row ← getRawDataLine(*line*);
> > result ← fft(row) × conjugate(fft(RangeChirp));
> > saveResultRow(*line*, ifft(result));
>
> **end**

The process is applied in the frequency domain, so that the previously defined *RangeChirp* $v(\tau)$ is transformed via the Fast Fourier Transformation (FFT). Its complex conjugate is multiplied with the Fourier-transformed raw data line. The result is inverted with the inverse of the Fast Fourier Transformation (IFFT) and saved.

Interestingly, although SAR processing is very different in azimuth and range direction, focusing in azimuth is rather similar to the range focusing. Figure 2.8 shows an example of the real part of an azimuth chrip. Again, a reference function is formed in azimuth with

$$v(t) = \exp(i \cdot \pi \cdot K_a \cdot t^2), \quad -\frac{t_a}{2} < t < \frac{t_a}{2}$$

where K_a is the frequency modulation rate in azimuth (FM rate) and t_a is the aperture time. $v(t)$ is formed with a step width Δt of

$$\Delta t = \frac{1}{\text{PRF}}$$

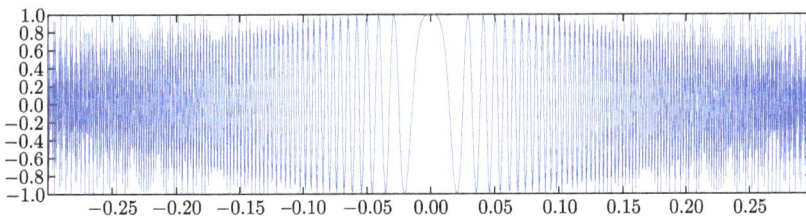

Figure 2.8   Real part of the azimuth chirp.

where PRF is the pulse repetition frequency. The FM rate K_a can be calculated from the effective velocity of the sensor v_{sensor} and the range to the target r with

$$K_a = -2 \frac{v_{sensor}^2}{\lambda r}$$

In pseudo-code, the focusing in azimuth can be described as

2.8. Azimuth focusing:
    input : Range compressed raw data
    output: Focused SAR image
    for *column* ← 0 **to** *size_{image}* **do**

```
col ← getRawDataColumn(column);
result ← fft(col) × conjugate(fft(AzimuthChirp));
saveResultColumn(column, ifft(result));
```

    end

The process is similar to range focusing, with the *AzimuthChirp* being $v(t)$. The range reference function is based on the waveform of the transmitted chirp only. The azimuth reference function depends on the geometry and is also adapted to the range.

One effect in SAR raw data processing is the Range Cell Migration (RCM), which is an effect based on the distance change between any fixed point and the radar along its trajectory. RCM can cause azimuth defocusing and needs to be corrected. As RCM is a two-dimensional and space-variant problem, it is the most challenging aspect of SAR data focusing. Several approaches for this have been developed over time. The most commonly known are those based on $\omega-k$ processors (Cafforio *et al.*, 1991), Range-Doppler algorithms (Jin & Wu, 1984), as well as chirp scaling approaches (Raney *et al.*, 1994). Very detailed analyses and comparisons of these processors, as well as more information on the focusing process can be found in several books (Carrara *et al.*, 1995; Cumming & Wong, 2005; Curlander, 1982).

2.4   Establishing a Target's Position with SAR

SAR systems can be precise geodetic measurement instruments. However, the SAR system measures only the distance between the antenna and the backscattering object precisely. Nevertheless, we could argue that the precise determination of time difference between each pixel in azimuth direction is also an intrinsic feature of the SAR system.

The geometrical properties of an SAR image are therefore quite different from optical imagery. SAR geometry is based on time, with the traveling time of the signal in range direction, the so-called fast-time, and the acquisition time in azimuth, the so-called slow-time. This leads to several effects in the SAR geometry.

2.4.1   SAR Geometry in Range Direction

The geometry in the range direction depends on the running time of the signal, which depends on the distance between the target and the sensor. The position is ambiguous though, as the possible positions can be described by a circle centered at the sensor's position with a radius of range r, assuming that the position of the sensor in azimuth direction is precisely known, otherwise we would need to assume a sphere. Taking the looking direction of the sensor into account, the circle describing the possible positions of a radar echo in space can be reduced to an arc. A position in the ground-range direction, described as x in Figure 2.9, can be established when intersecting this arc with an estimated or known height. Using a constant estimated height, e.g. 0, for the whole image, a ground-range image can be generated. Alternatively, a known DEM can be used. The ground range position x can be derived from

$$x = \sqrt{r^2 - (H - h)^2} \tag{2.20}$$

where H is the height of the sensor above the plane describing the ground and h is the height of the target with respect to that plane.

This works well as long as each target is at the height of the estimated ground plane. This is not the case in real images, leading to various distortions with respect to the topography, i.e. the height of the targets.

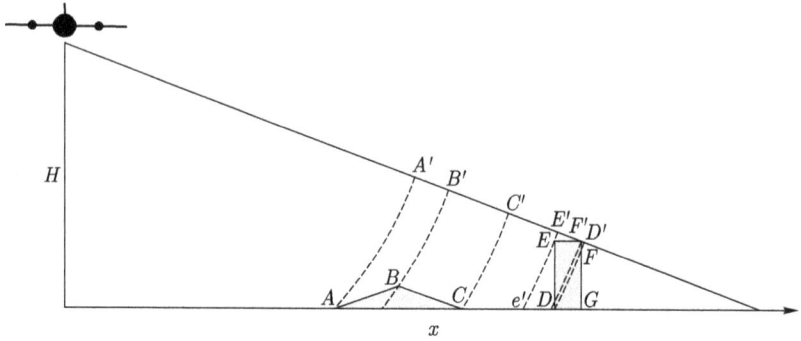

Figure 2.9   SAR imaging geometry in range direction.

These effects can be described with the help of Figure 2.9. The positions of various targets in that example are given with letters from $A{\sim}G$. Their respective positions in the so-called slant-range SAR image are shown with $A'{\sim}G'$. This position is determined by the intersection of an arc with the range distance r between the target and the sensor and the slant-range plane. Similarly, the so-called ground-range position of a target in the SAR image is determined by the intersection of that arc with the virtual plane describing the ground, often established at a height 0, but sometimes also at an estimated average height of the respective SAR image.

The targets residing at the height of the plane A, C, and D are projected at their correct ground-range position in the SAR imaging process, with their relative distances also being accurate. However, target B is elevated and is therefore closer to the sensor. Due to the position of B being determined by the intersection of the arc with range r and the slant-range or ground-range planes, B is positioned closer to the sensor in x, not representing the target's true position in x. As a result, the distance between A' and B' gets reduced, the so-called fore-shortening effect. Similarly, the distance between B' and C' gets enlarged in the SAR image.

The situation is more extreme for E and F, where we can find E' and F' appearing before D' in the SAR image. This effect is called layover and appears frequently in SAR images of buildings, where especially high-rise buildings might show a significant layover effect. Additionally, layover effects cause ambiguities in the SAR image, as, for example, the radar return of E will arrive at the same time as the radar echo of possible targets at the position marked with e' in Figure 2.9, or any other possible target on the arc of similar distance r for that matter.

Finally, there is no G' to be found in the SAR images. This is because G is not visible from the SAR system as G is located in the so-called radar-shadow. Therefore, G is not scattering any signal back to the sensor.

2.4.2 Geometrical Model of SAR Positioning

The radar coordinate system of an SAR image describes the travel-time of the signal in range direction, corresponding to the distance between the target and the sensor r, in one axis of the image. On the other axis, time elapsed since a given reference time is given. This time difference Δt can be used to establish the precise focused time of each pixel in the SAR image. Thus, for each pixel we have a precise information of the acquisition time of that pixel and the travel time of the signal from that sensor to the target and back.

Based on the azimuth time, the position of the sensor in space can be determined. This requires information on the sensor positions in space and time to be provided with the SAR image, which is an integral part of each SAR image's meta information. These orbit ephemerides are generally provided in a geocentric position based on the center of the Earth, as shown in Figure 2.10. Based on an ellipsoidal model of the Earth, e.g. WGS84, the position on the Earth's surface can then be determined based on these geocentric coordinates. The orbit ephemerides of the sensor S are typically provided with position information, velocity information, and sometimes also acceleration information, with a time interval between them.

As this time interval is typically much larger than the resolution of the SAR system in azimuth direction, interpolation of the position is required, assisted by velocity and acceleration information.

Based on this information, for a point on the Earth, the corresponding position in the SAR image can be derived. Keep in mind, that only the

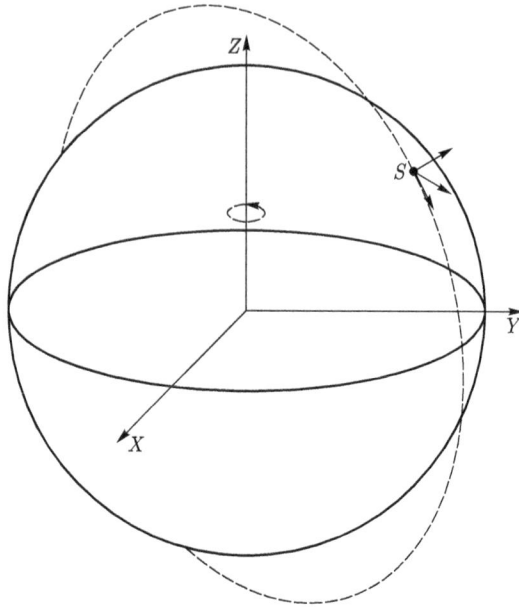

Figure 2.10   Position of satellite *S* in the geocentric coordinate system.

transformation of a given three-dimensional coordinate on the Earth, i.e. given the latitude, longitude, and height, the transformation to two-dimensional radar coordinates is valid. The back-transformation from two-dimensional radar coordinates to world coordinates is ambiguous. The transformation from world coordinates is normally implemented via geocentric coordinates and an iterative solution in the Range-Doppler process as described by Curlander (1982).

2.4.3   Geometrical Distortions in SAR Imaging

The most common geometrical distortions in range direction have already been discussed in Section 2.4.1 of Chapter 2. SAR imaging can be very precise, if the various distortions are corrected properly, which will be described in more detail in Chapter 10. However, in addition to the effects in range directions due to the running-time geometry of the SAR image, there are also distortions caused by the SAR imaging process, i.e. by the formation of the synthetic aperture. The formation of the synthetic antenna over the flight-path of the sensor is a process that takes time, so an SAR image is a combination of various radar data acquisitions over time. Moving targets are not considered in this process and will therefore be imaged incorrectly. We can understand this effect either by considering SAR as the formation of a synthetic antenna over time, or as a Doppler-beam focusing system. With the position of a target being determined by its zero-Doppler position, it becomes evident that a target producing a Doppler effect by its own motion will become displaced in the azimuth direction.

The specific way of the geometrical distortions of moving targets, i.e. displaced or defocused, depends on the target's velocity relative to the radar sensor (Jao, 2001). A target that is moving along-track to the sensor, i.e. along the same path as the radar sensor, appears smeared in the SAR image. Across-track movement, i.e. the target moving toward or away from the sensor in range direction, means the target motion producing a shift in the target image position in the azimuth direction.

Assuming that the object moves across-track with constant velocity in range direction v_{range}, then the variation of the range history is linear causing a secondary linear phase trend in the signal. By following the Fourier transformation law, the linear phase element correlates to a time shift Δt in the time domain, with

$$\Delta t = \frac{2v_{\text{los}}}{\lambda \text{FM}} \tag{2.21}$$

where FM is the frequency modulation and v_{los} is the velocity in the line-of-sight direction of the satellite. v_{los} is related to the velocity of the target v_{range} in range direction with

$$v_{\text{los}} = v_{\text{range}} \sin \theta_{\text{inc}} \tag{2.22}$$

The displacement in azimuth direction Δ_{az} is (Weihing *et al.*, 2006)

$$\Delta_{az} = -r \frac{v_{\text{los}}}{v_{\text{sensor}}} \tag{2.23}$$

where v_{sensor} is the velocity of the sensor.

2.5   Speckle Effect

SAR images, as all coherent imaging systems, suffer from the so-called speckle effect (Figure 2.11). The coherent sum of the amplitudes and phases of scatterers with random distribution on a resolution cell causes strong fluctuations of the backscattering from a resolution cell. For distributed scattering, where many scatterers contribute significantly to the backscattering of a resolution cell, the intensity and phase are therefore no longer deterministic. The total complex reflectivity of a resolution cell is given by

$$\Phi = \sum_i \sqrt{\sigma_i} e^{j\varphi_i} e^{-j\frac{4\pi}{\lambda}r_i} \tag{2.24}$$

where i is the number of scatterers in the resolution cell. Speckle is often described as noise, but it is a physical measurement of the distribution of scatterers in the resolution cell. Nevertheless, for many applications, speckle acts as noise when processing an SAR image.

Speckle can be reduced by multi-looking. Multi-looking is a non-coherent averaging of the image often used to improve interpretability. Spatial multi-looking implemented using a moving

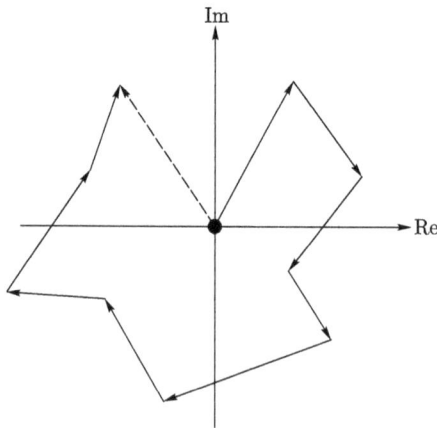

Figure 2.11   Speckle pattern of an electromagnetic wave.

window reduces the spatial resolution of the image. As shown in Figure 2.12, having a stack of SAR images, temporal multi-looking is possible, assuming that there are no relevant changes on the ground.

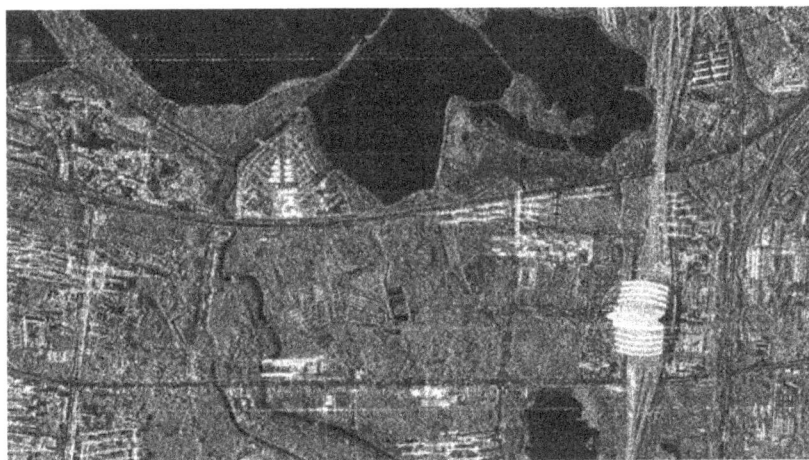

(a)

(b)

Figure 2.12 TerraSAR-X stripmap image of Wuhan railway station; (a) single-look amplitude image; (b) mean amplitude image of 282 images. © DLR - 2011–2019. All Rights Reserved.

2.5.1   Speckle Statistics and Multi-Looking

From one resolution cell, many contributions arrive with different amplitude and phase. The real and imaginary parts are independent and Gaussian-distributed. Their joint probability distribution function (PDF) describes the probability of any particular real and imaginary value occurring as a result of the combined contribution of the various contributions coming from that particular resolution cell.

Mathematically, this can be described as (Oliver & Quegan, 2004)

$$P(z_1, z_2) = \frac{1}{\pi m} \exp -\frac{z_1^2 + z_2^2}{m} \qquad (2.25)$$

where z_1 and z_2 are the real and imaginary components of the wave, $P(z_1, z_2)$ is the probability of the resulting signal that has these real and imaginary components and the πm is for normalizing the PDF. We expect the phase angles to be uniformly distributed. The PDF of the amplitude is independent of the phase and is given by the Rayleigh distribution.

With the speckle effect being very strong, the radar cross section of a target suffering from the speckle effect can only be successfully estimated by combining a number of measurements and averaging them. This can, for example, be done with a smoothing filter that averages the intensity of the N pixels in the averaging window. The larger N, the better, in theory, is the estimation of the intensity. However, the averaging is only valid if it includes only the pixel of the same target. If there are different targets inside the averaging window, the estimated average is not valid, as it is derived from different objects with a different radar cross section. Therefore, the averaging window and N should not be too large to ensure the homogeneity of the objects inside the window.

SAR offers a unique way of providing independent samples during the acquisition of the SAR image itself. In this process, the synthetic aperture generated along azimuth in the SAR imaging process is split into several sub-apertures, as shown in Figure 2.13. If, for example, four sub-apertures are formed, we would say that we generated four looks. The higher the number of looks N, the lower the spatial resolution

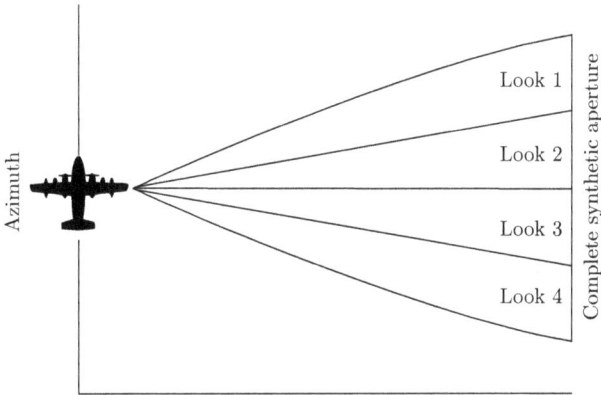

Figure 2.13 Forming a multi-look image from four sub-apertures.

in azimuth gets. These looks can be averaged to generate a multi-look image. Different from the averaging approach, each look corresponds to a measurement of the same resolution cell on the ground, although there is a larger resolution cell due to the reduced spatial resolution.

In SAR image processing, both approaches are often referred to as looks and multi-looking. In both cases, the variance of the intensity I values gets smaller with an increase in N:

$$\text{var}(I) = \frac{\bar{I}^2}{N} \tag{2.26}$$

This relationship can also be used to derive the so-called equivalent number of looks N_e with

$$N_e = \frac{\bar{I}^2}{\text{var}(I)} \tag{2.27}$$

This can be useful if the number of looks is unknown, which can also be due to filtering operations or to validate the effectiveness of the used multi-looking approach.

2.5.2   Speckle Filtering

In order to decrease the speckle noise, incoherent averaging can be applied through specialized speckle filters. Typically, these filters also work on an averaging window, where it is important that homogenous areas are filtered. A variety of speckle filters have been developed over the years with different assumptions on the underlying probability distribution functions. Very common examples are the Lee-Sigma Filter (Lee, 1983), Frost Filter (Frost *et al.*, 1982), and Gamma-MAP Filter (Lopes *et al.*, 1990). However, there is a large variety of other speckle filters that have been developed over time.

These filters are based on a simple moving box window and it is difficult to ensure homogeneity inside the window. Newer approaches therefore tend to use adaptive filtering windows that ensure a certain homogeneity within a window. One of these approaches is, for example, DespecKS (Ferretti *et al.*, 2011), which will be discussed in more detail in Section 8.3 of Chapter 8. Another approach is non-local filtering (Zhu *et al.*, 2018). Non-local techniques can reduce noise while preserving finer structures. This is achieved by a weighted averaging of similar pixels. The keyword is similarity, which in local filtering is estimated by spatial features, i.e. pixels next to each other are likely to be similar. In non-local filtering, two pixels are considered similar if the surrounding image chips are similar in appearance or statistical homogeneity. Similarity in this context is often defined as the Euclidean distance between the vectors of intensity.

2.6   SAR Satellite System for Ship Detection

In our example, we designed an SAR satellite system. We wanted to achieve a 10 m resolution in range and azimuth to clearly identify smaller ships. The radar will operate in C-band with a wavelength of 5.15 cm. We assume a range of 800 km from the platform to our targets. The platform has a velocity of 7500 m $\cdot$ s^{-1}.

2.6.1 Signal-to-Noise Ratio

Our first consideration is the signal-to-noise ratio (SNR). A significant source of noise in an SAR system is thermal noise from the system itself. The power of the noise P_n can be derived from the Boltzmann constant k (1.38×10^{-23} J $\cdot$ K^{-1}), the noise temperature T_n and the systems bandwidth B:

$$P_n = k \cdot T_n \cdot B$$

We do not consider other sources of noise at the moment, as they are more likely to appear at the edges of the microwave spectrum. The SNR is a measurement of the signal quality and is the ratio of the received power of the system P_r and the power of the noise P_n:

$$\text{SNR} = \frac{P_r}{P_n}$$

So, for our imaging radar system, we can derive it from

$$\text{SNR} = \frac{P_r}{P_n} = \frac{A_e G P_t \sigma^0 A}{(4\pi)^2 r^4 k T_n B} \tag{2.28}$$

There are many variables to consider, which influence each other, e.g. the resolution of the system depends on the antenna size A_e and the bandwidth B. So, for the resolution, we want to have small antenna sizes and high bandwidth, while when keeping a high SNR, the opposite is desired.

In our system, we want to keep an SNR > 30 dB and the maximum power we can transmit P_t with 5000 W. The next step would be to determine the necessary bandwidth we need to reach 10 m resolution in the slant range.

$$\delta_{rg} = \frac{c}{2B} \tag{2.29}$$

so

$$B = \frac{c}{2\delta_{rg}} \tag{2.30}$$

Therefore, $B \approx 15$ MHz for a desired 10 m range resolution.

With an antenna gain G of 40 000, a noise temperature T_n of 100 K, and a desired footprint of 60 km^2 and an assumed σ^0 of 4%, we get an SNR of 25 dB if we design an antenna with an area A_e of 10 m^2. This is close to our desired 30 dB, but not quite right. So, if we decide on an antenna size of 20 m^2 and a respective increase of G to 80 000, we achieve an SNR of 31.6 dB sufficient for our needs.

We plan to design the antenna with 10 m length in azimuth and 2 m height. So, the theoretically achievable resolution of the synthetic aperture in azimuth is δ_{sa}:

$$\delta_{sa} = \frac{l_{ra}}{2} = 5 \text{ m}$$

which exceeds our specifications. Such an antenna will have a footprint of

$$2 \cdot r \cdot \tan \frac{\lambda}{l_{ra}} \tag{2.31}$$

which corresponds to 8.2 km in azimuth (10 m length of the antenna) and 41 km across track (2 m height of the antenna). So, our SAR system will have a swath width of 41 km. Now, our footprint is actually larger with 336 km^2, further increasing the SNR.

2.6.2   Pulse Repetition Frequency

To form a complete synthetic aperture and avoid aliasing longer wavelengths back into shorter wavelength, a radar pulse needs to be sent at a distance of half of the antenna length l_{ra} or shorter. The PRF needs to be larger than

$$\text{PRF} > \frac{2v_{\text{sensor}}}{l_{\text{ra}}}$$

which in our example corresponds to

$$\text{PRF} > \frac{2 \times 7500 \text{ m} \cdot \text{s}^{-1}}{10 \text{ m}} = 1500 \text{ Hz}$$

However, the PRF can also not be too large to avoid ambiguities between the pulses from near range and far range. So, the PRF needs to be smaller than

$$\text{PRF} < \frac{1}{t_{\text{far}} - t_{\text{near}}}$$

Recalling that the real antenna has an angular resolution of

$$\alpha = 2 \tan \frac{\lambda}{l_{\text{ra}}}$$

corresponding roughly to 3°. Now, let's assume an incidence angle $\theta_{\text{inc}} = 45°$, this will put our sensor at a height of approximately 566 km. Keep in mind that this is a simplification, as we did not account for the curvature of the Earth.

Staying with these simplified results, we can calculate the near-range and far-range distances and the time difference Δt between them, which is 0.14 ms, corresponding to a maximum PRF of 3574 Hz. However, this is dependent on the incidence angle. Nevertheless, we have a wide range of PRF and can pretty safely select a PRF with

$$1500 \text{ Hz} < \text{PRF} < 3574 \text{ Hz}$$

2.6.3   Satellite Orbits and Repeat Time

After designing our sensor, we are happy with ourselves until we join a first meeting with possible users of the data. Suddenly, as it is often the case, new demands on the system are raised. The users are concerned that the system is not getting data at a high enough temporal resolution. To monitor the ships, they would like to have a daily update on the maritime situation. We were lucky and talked them out of it, but they insisted on a weekly data coverage.

To ensure global coverage, satellites are typically launched into a Sun-synchronous orbit (Figure 2.14) and our satellite is no exception. The satellite orbits around the Earth, while the Earth rotates ensuring that over time each area is covered.

In the introduction, we stated that one of the big advantages of the SAR system is its independence from the Sun as light source. So, why are SAR satellite systems following a Sun-synchronous orbit? The SAR system itself can operate at day and night, but a satellite system needs the Sun to provide the energy for the solar panels. So, a spaceborne sensor is in fact not independent of the Sun.

In the following, we roughly estimate how long it would take the satellite to cover the globe. However, keep again in mind that this is a very simplified calculation and many more parameters need to be taken into consideration when designing a satellite orbit and calculating the repeat time of a satellite.

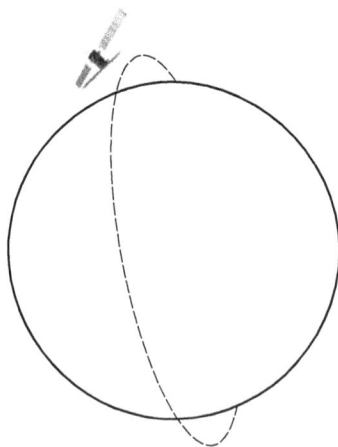

Figure 2.14   Satellite orbiting in a Sun-synchrounous orbit.

Assuming an Earth radius of 6378 km plus the satellite height of 566 km, our satellite has to travel 43 630 km in one orbit, taking approximately 1.6 hours with the assumed velocity v_{sensor} of 7.5 km $\cdot$ s^{-1} and a swath width of 41 km, we can cover approximately 40 000 km around the equator of 977 images at 41 km each. So, it will take us approximately 65 days to come back to the same place on the Earth, reducing the temporal resolution of our sensor to 65 days. This, unfortunately, is below the specifications of our users.

In satellite remote sensing, however, we have the opportunity to see the same area two times though: once from an ascending orbit and once from a descending orbit. For our application, detecting the ships would be possible once from the ascending orbit and once from the descending orbit, reducing our time to 32 days (Figure 2.15).

This is better, but still not close enough to the user requirement for weekly updates. One way is to redesign the satellite with a stronger focus on a larger swath width, starting with decreasing the antenna height. This would decrease our SNR but would also mean we would need to redesign everything. If the project went far along, it would probably mean that the complete system, platform and launch vehicle needs to be designed again, which is not something one would like to do. Fortunately, the SAR system can be designed as a very flexible remote

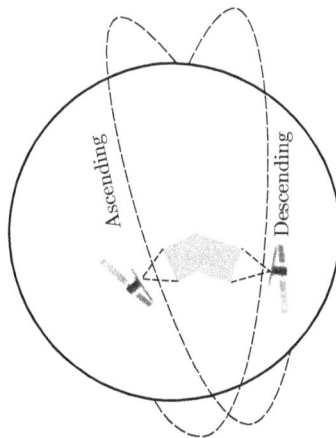

Figure 2.15 Ascending and descending orbit.

sensing system because it can be operated in different SAR acquisition modes.

2.6.4   SAR Acquisition Modes

2.6.4.1   *Stripmap mode*

In our previous example, the SAR system was operated in the so-called stripmap mode, which is the standard mode of operation for most SAR systems and works with fixed antennae. Antennae allowing for a flexible adjustment of the looking direction during image acquisition can be used in additional acquisition modes. The adjustment of the looking direction of an antenna is, in most systems, realized by using a phase array antenna that can programmatically switch looking directions up to a certain degree. It is also possible to steer the antenna direction mechanically, which allows an even larger degree of look-direction changes, however, this causes oscillation of the system and is therefore not recommended for systems depending on very precise antenna patterns, as we, for example, need for SAR interferometry.

2.6.4.2   *Spotlight mode*

By adjusting the looking direction of the sensor in azimuth during acquisition, we can enlarge the synthetic aperture that is formed and therefore increase the spatial resolution of the system in azimuth (Figure 2.16). However, the area of image acquisition along the track of the sensor is also reduced.

Example 2.9   In our example, the length of the synthetic aperture is

$$l_{sa} = 2\frac{\lambda}{l_{ra}}r = 2 \times \frac{0.0515 \text{ m} \times 800\,000 \text{ m}}{10 \text{ m}} = 8240 \text{ m}$$

Using this in our basic radar resolution equation we get the previously mentioned 5 m resolution in azimuth:

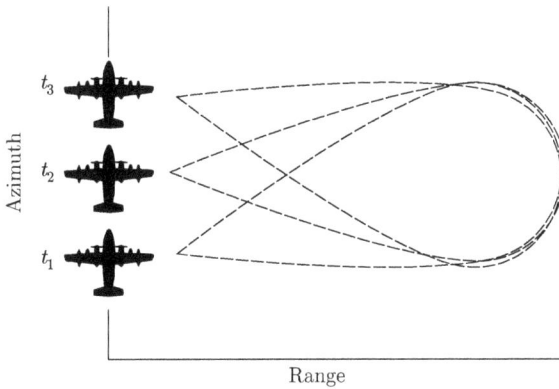

Figure 2.16 Spotlight SAR.

$$\delta_{az} = r\frac{\lambda}{l_{sa}} = 5 \text{ m} \qquad \square$$

For spotlight mode, we can shift the antenna looking angle in azimuth, the so-called azimuth antenna steering spin $\Delta\Theta_s t$ is typically about 1°, but can also be even larger (Kraus *et al.*, 2016).

Assuming 1°, our synthetic antenna gets significantly longer, by approximately 14 km, reducing the azimuth resolution to about 1.86 m. To ensure complete sampling, this requires an increase in PRF in our example of up to 3763 Hz.

However, this is above the limit of 3574 Hz as previously mentioned to avoid ambiguities from near- and far-range signals. We can limit the extent of the swath width to counter that effect or limit the increase in resolution to fit into the limit. For example, if we decide to limit the PRF to 3400 Hz, we could limit our resolution to about 2.1 m, steering the antenna with $\Delta\Theta_s t = 0.8°$ and limiting the PRF to 3400 Hz. Keep in mind that this is dependent on the incidence angle. In our simplified example, we used a fixed incidence angle $\theta_{inc} = 45°$.

2.6.4.3 ScanSAR mode

Increasing the resolution with the spotlight mode is useful in many applications. But this would not solve our problem of limited temporal

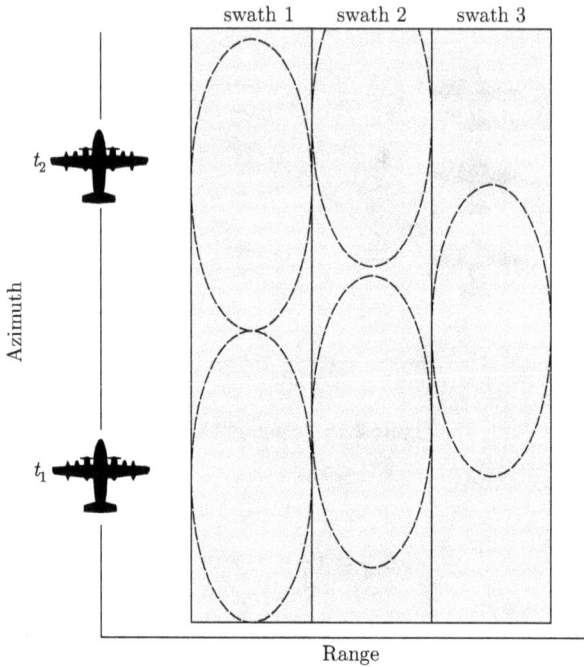

Figure 2.17   ScanSAR mode.

resolution. Actually, it would worsen the problem because in spotlight mode, the coverage along track direction is limited, so that our temporal resolution would even decrease.

Having a flexible antenna in incidence θ_{inc} allows us to increase our swath width using the ScanSAR mode. With an increased swath width, we can have larger coverage and increase our temporal resolution (Figure 2.17).

In ScanSAR mode, the acquisition time is divided by the number of scans. Each scan is acquiring data with a different incidence angle θ_{inc}, covering a much larger area in range. However, for each scan, the size of the synthetic antenna is reduced accordingly, leading to a reduction of the resolution in azimuth. Furthermore, the bandwidth needs to be divided between the scans, leading to a similar reduction in the range resolution.

So, if we assume three scans and accordingly increase our swath width three times from 41 km to 123 km, we decrease the synthetic antenna by one-third to 2747 m to get an azimuth resolution of about 15 m. Actually, there would be a slight overlap between the scans, so that we can assume a swath of probably around 120 km. There is acquisition time lost during the switch from one scan to another, further decreasing the available synthetic aperture length, so that the achievable resolution in azimuth will be above the theoretical value of 15 m.

With an increased swath of 120 km, we can assume that we need 334 images to cover the 40 000 km around the equator, so we could reduce the time for a repeat orbit acquisition to about 22 days. Including ascending and descending acquisitions, 11 days are possible, which is below the requirements that have been discussed before, but still much closer. However, this comes at the cost of reduced resolution in azimuth of 15 m and in range of about 30 m, which is below our previous specification of 10 m resolution.

2.6.4.4   TOPS mode

A problem of the ScanSAR mode is the time the sensor needs to switch from one scan to the next. During this time, no data is acquired, which can lead to scalloping effects in the image. This is an undesired side effect of the ScanSAR mode, but is improved by the TOPS mode.

TOPS is the opposite of Spot and it is a combination of an inverse spotlight mode and the ScanSAR mode. In TOPS, the antenna is shifted in azimuth, however opposite to the spotlight mode. In TOPS, for each burst, the antenna starts looking backward and then gradually changes the looking direction until looking forward. This reduces the synthetic antenna size and therefore reduces the resolution in azimuth even further. It increases the area covered in azimuth though and creates overlaps between each of these sub-acquisitions, the so-called bursts. Similarly, TOPS achieves an overlap between the scans in range, so in the border areas between bursts and swaths, there are large overlapping areas, as shown in Figure 2.18.

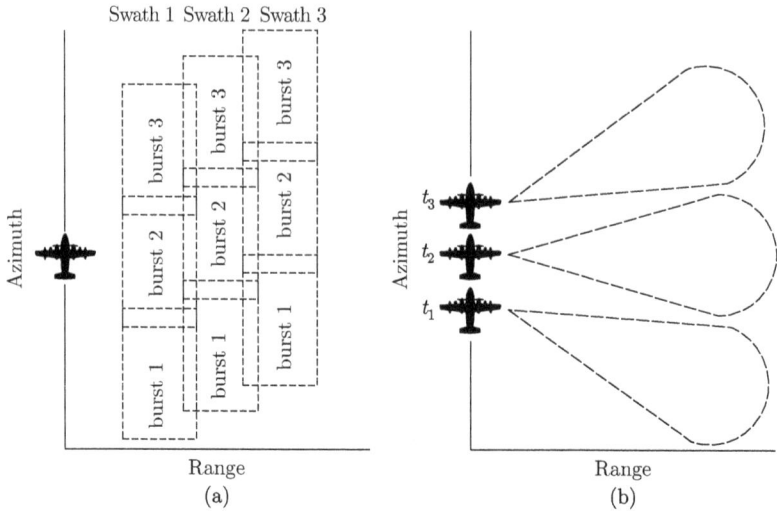

Figure 2.18   TOPS mode acquisition: (a) burst and swath overlap; (b) single burst acquisition with beam sweeping from backward looking to forward looking for each single burst.

This overlap ping area can be used to correct for the scalloping effect. The TOPS mode on the other hand is a rather complex mode and causes several problems during SAR image processing, which we will discuss later on. This mode is, unlike ScanSAR, widely used; the Sentinel-1 mission is using TOPS as the standard acquisition mode (Potin *et al.*, 2018), making it an important means for SAR acquisition, allowing for the global availability of the Sentinel-1 data from the Copernicus program.

2.7   SAR Polarimetry

An electromagnetic wave, as every transverse wave, oscillates perpendicular to the direction of the wave. Polarization is the geometrical orientation of the oscillations. For a vertically polarized wave, the electric field oscillates vertically, while the magnetic field oscillates perpendicularly to the electric field, as shown in Figure 2.19.

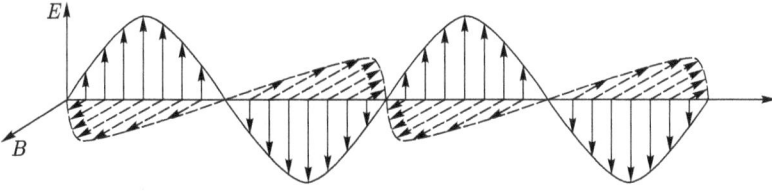

Figure 2.19 Polarization of an electromagnetic wave.

SAR polarimetry is used to derive physical information from the measurement of the polarimetric scattering behavior of scatterers. SAR polarimetry uses a 2 × 2 complex scattering matrix to describe the transformation of a two-dimensional transmitted wave vector E^t into the received wave vector E^r:

$$\begin{bmatrix} E_H^r \\ E_V^r \end{bmatrix} = \frac{e^{-jkr_0}}{r_0} \begin{bmatrix} S_{HH} & S_{HV} \\ S_{VH} & S_{VV} \end{bmatrix} \begin{bmatrix} E_H^t \\ E_V^t \end{bmatrix}^* \tag{2.32}$$

where k is the wavenumber with $k = 2\pi/\lambda$.

The scattering matrix is measured by transmitting two orthogonal polarizations, typically horizontal (H) and vertical (V), and receiving two orthogonal polarizations. This is realized by switching the transmitted and received polarization between each pulse, which in turn reduces the spatial resolution for each of the measured polarizations.

In SAR polarimetry the first letter indicates the transmitted pulse and the received pulse is given by the second letter, so that S_{HH} is the part of the scattering matrix with a horizontally transmitted and horizontally received wave, whereas S_{VH} is vertically transmitted and horizontally received. In the mono-static case, the scattering matrix is symmetric with $S_{HV} = S_{VH}$, which is then often expressed as S_{XX}.

The scattering matrix $[S]$ describes the scattering behavior of a single point-like scatterer, but fails to fully express the scattering of distributed targets. This requires a second-order statistical description to incorporate the quasi-randomness found in speckling.

An often used descriptor for SAR polarimetry is the Pauli vector $\mathbf{k}_p$. The Pauli vector is often used for visualizations of fully polarized SAR images, as shown in an example in Figure 2.20.

Figure 2.20   Polarized TerraSAR-X image near Mount Tai shan, China, in Pauli visualization (© DLR, 2010).

$$\boldsymbol{k}_{\mathrm{p}} = \frac{1}{\sqrt{2}} \begin{bmatrix} S_{\mathrm{HH}} + S_{\mathrm{VV}} \\ S_{\mathrm{HH}} - S_{\mathrm{VV}} \\ 2S_{\mathrm{XX}} \end{bmatrix} \qquad (2.33)$$

Scattering decomposition approaches separate the polarimetric backscattering of distributed scatterers permitting interpretation of the scattering process. The basic scattering mechanisms are surface, dihedral, and volume scattering. Separating these scattering components can be used for classification, segmentation, or pre-processing steps. SAR polarimetry is a very wide field, only the surface was scratched here. A more complete description of SAR polarimetry can be found, for example, in Lee and Pottier (2009).

2.8   Conclusions on the Developed SAR System for Ship Detection

To be honest, we would have to conclude that we cannot be too happy with the system we designed. It did not actually meet our specifications and we would probably need to rethink the design. Considering the speckle effect, we should probably even consider increasing the resolution of the system further at least in one direction to have a multi-looking resolution of 10 m, which will be very useful for a detailed ship detection of smaller vessels.

Nevertheless, as stated at the beginning, this chapter does not cover SAR system design, nor is it really about ship detection. It is an introduction to SAR systems to give an insight into SAR systems and the contradictions and considerations for SAR system design. No SAR system can satisfy all needs, and therefore, a system should be designed according to specific needs. Similarly, the data requirements should be based on the application and not all SAR systems are suitable for each individual task.

Modern methods of ship detection from SAR images have not been discussed at all in this chapter. We also discussed detecting ships as hard targets with high backscattering, whereas many ship detection approaches identify ships by their wakes (Eldhuset, 2004; Kuo & Chen, 2003). Identifying the ships by their high RCS still requires the selection of a threshold. The Constant False Alarm Rate (CFAR) algorithm is widely used for computing an adaptive threshold (Crisp, 2004). Approaches based on machine learning are becoming more commonly used as well, so that overall the field of ship detection in SAR images continues to be a vibrant research field.

Part II   DEM Generation

Chapter 3

SAR Interferometry

The creation of a digital elevation model (DEM) is essential for topographic mapping, and DEMs constitute fundamental basic data required for many geospatial applications. The specification for each DEM differ for different applications. While some applications, like, for example, classical topographic map generation, have a high demand for the absolute correctness of the height values, other applications, for example, in hydrology, have high demands for relative accuracy, which is the relative accuracy from one data point to another data point in the DEM. For hydrology, absolute errors matter less than keeping the relative error small enough to, for example, ensure the correctness of river flow modeling.

There are various methods for DEM generation available, starting from traditional surveying, which offers high accuracy, but is labor-intensive and therefore quite costly if large areas need to be covered. Airborne photogrammetry offers excellent precision and is relatively cheap and widely used (Haala & Rothermel, 2012). LIDAR offers very high precision, but due to the limited footprint of the system is more expensive and more suitable for smaller areas (Jaboyedoff *et al.*, 2010; Wang, 2013). SAR offers larger footprints at high precision, and is often the data source of choice for regional and global DEM projects (Krieger *et al.*, 2007). The ability to acquire data under (almost) all weather conditions makes it also more cost-effective in terms of mission planning and airplane or satellite capacity management.

3.1   Introduction to SAR Interferometry

SAR interferometry uses the phase difference between two coherent radar signals. Differences in phase can be considered as differences in distance between two acquisitions. These small differences can then be used to infer height differences between pixels, from which a DEM can be derived.

The basic methodologies for interferometric SAR were presented at the end of the 1980s. Gabriel and Goldstein (1988) presented the first SAR interferometry data based on the SIR-B mission. One year later, they presented the first differential interferometry results (Gabriel *et al.*, 1989). The applicability of D-InSAR for seismic applications was demonstrated using ERS data (Massonnet *et al.*, 1993). The probably most well-known interferometric mission was the Shuttle Radar Topography Mission (SRTM), generating the widely used SRTM DEM (Farr *et al.*, 2007). More recently, the TanDEM-X mission created a global DEM based on bi-static interferometry at unprecedented precision (Krieger *et al.*, 2013).

3.1.1   Amplitude and Phase Components of SAR Images

Every pixel of an SAR image contains two pieces of information, the amplitude and the phase, as shown in Figure 3.1. This information is typically represented as a complex number containing a real value and an imaginary value. In the previous chapters, we ignored the phase and only analyzed the amplitude values of SAR images. While the amplitude represents the backscattering intensity, the phase represents the distance between the sensor and an object. However, the phase is limited to values between $-\pi$ and π. We can understand the phase as going around a circle with 360°, starting again at 0°. We call this a wrapped phase. The phase information is therefore ambiguous.

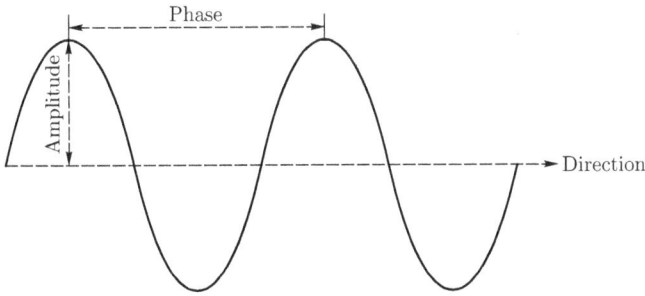

Figure 3.1 Properties of an electromagnetic wave.

With radar systems, we can measure the phase of the backscattered signal. However, this gives us only a rather small part of the information about the overall distance. The signal travels from the sensor to a backscattering object and back. This is the overall distance $2r$, where r is the distance between the sensor and the object. While the electromagnetic wave travels along the signal path, many full phase cycles are completed on the path. The phase of the signal ϕ can be calculated with

$$\phi = -\frac{4\pi}{\lambda}r + \phi_{\text{scatt}} \tag{3.1}$$

where λ is the wavelength and ϕ_{scatt} is the phase contribution from the scattering. However, the actually measured phase is not ϕ, but the wrapped phase φ, where

$$\varphi = W\{\phi\} \tag{3.2}$$

where $W\{\}$ is the wrapping operator and φ is wrapped between $-\pi < \varphi < \pi$.

Unfortunately, φ is the much less useful information because from ϕ, we could derive the distance precisely, while with only knowing φ, not much useful information is to be gained from a single image.

Figure 3.2 shows clear amplitude texture, however, the phase image shown in Figure 3.3 looks like random noise.

Figure 3.2   TerraSAR-X amplitude image of Uluru acquired on 2009-02-12 (© DLR, 2009).

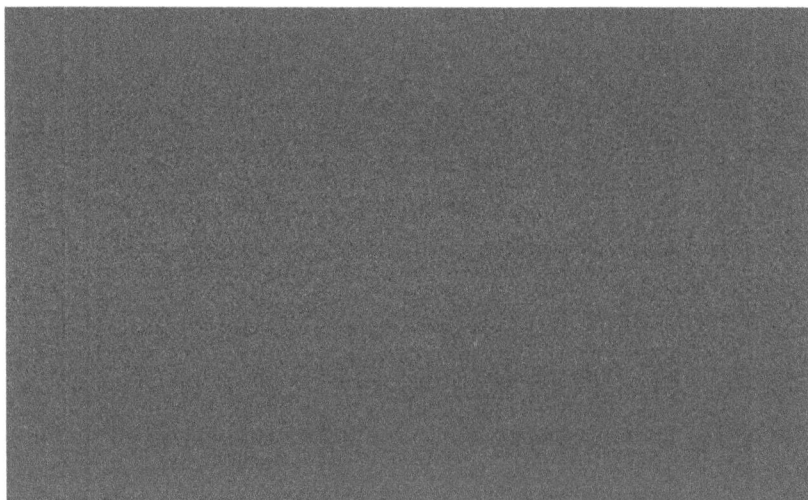

Figure 3.3   TerraSAR-X phase image of Uluru acquired on 2009-02-12 (© DLR, 2009).

3.1.2   Phase Differences as Basis for Height Determination

Although the phase information of a single SAR image is not very useful, the difference between the phases from two antennae can provide precise geodetic information. Several prerequisites have to be fulfilled. The radar signal needs to be coherent and the phase calibrated. That is to say, that not every available SAR system is usable for interferometry, but nowadays most systems are appropriate. It means, furthermore, that the distance between the two antennae should not be too big to keep the two phases coherent.

The requirement of having two antennae can be fulfilled by acquiring both signals at the same time in a so-called bi-static mode with one antenna transmitting and two antennae receiving. This is called single-pass interferometry. However, instead of using two antennae, one can also acquire two images from slightly different positions with the same antenna, but at different times. This is called repeat-pass interferometry.

In repeat-pass interferometry, changes on the ground or in the properties of the atmosphere between the acquisition will cause a so-called loss of coherence. Depending on the time interval, the wavelength, and the amount of changes, this loss of coherence can quickly lead to errors and a lack of information. In the following passages, we will assume repeat-pass interferometry.

R Please be aware that in some of the equations, there are slight differences between repeat-pass interferometry and bi-static interferometry.

Now, if we assume having two acquisitions from slightly different positions, as shown in Figure 3.4, we can start getting useful results. Based on Figure 3.4, we can define

$$\phi_1 = -\frac{4\pi}{\lambda}r + \phi_{scatt,\,1} \tag{3.3}$$

$$\phi_2 = -\frac{4\pi}{\lambda}(r + \Delta r) + \phi_{scatt,\,2} \tag{3.4}$$

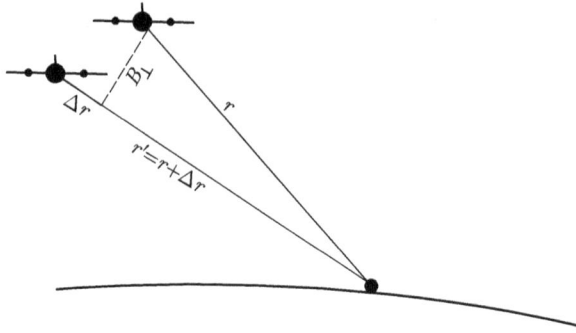

Figure 3.4   Interferometric geometry.

If we form the interferometric phase ϕ, i.e. the phase difference,

$$\phi = \phi_1 - \phi_2 = \frac{4\pi}{\lambda}\Delta r \tag{3.5}$$

3.1. Interferometric phase:

$$\phi = \frac{4\pi}{\lambda}\Delta r \tag{3.6}$$

This is valid in the mono-static case and only if $\phi_{scatt,1} = \phi_{scatt,2}$. If we can assure this, then the interferometric phase depends on the distance difference between the two acquisitions. However, in reality, we are dealing with wrapped phases, so that

$$W\{\phi\} = W\{\phi_1\} - W\{\phi_2\} \tag{3.7}$$

Here, the equation is in principle still valid, but Δr needs to remain small. As we will find out, phase wrapping is a major problem in SAR interferometry.

Differences in range can be caused by different effects. Differences in the topographic height of two points can cause range differences due to slight variations in the acquisition geometry between the SAR images,

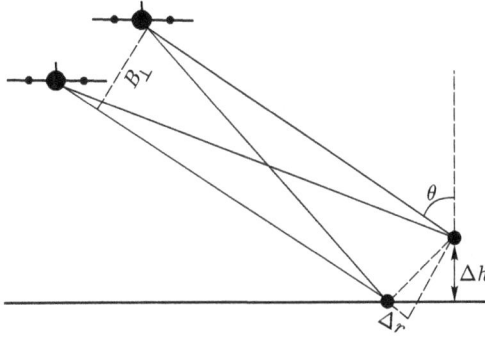

Figure 3.5 Relative interferometric phase.

allowing us to assess topographic height differences between points by analyzing the relative interferometric phase between two points.

In SAR interferometry, the measurements are taken on the relative interferometric phase, which is the phase difference between two pixels.

$$
\begin{aligned}
\Delta\phi &= \phi_{\text{pix}_1} - \phi_{\text{pix}_2} \\
&= \frac{4\pi}{\lambda}\Delta r_{\text{pix}_1} - \frac{4\pi}{\lambda}\Delta r_{\text{pix}_2} \\
&= \frac{4\pi}{\lambda}\left(\Delta r_{\text{pix}_1} - \Delta r_{\text{pix}_2}\right) \\
&= \frac{4\pi}{\lambda}\Delta r
\end{aligned}
\tag{3.8}
$$

As in the case of the interferometric phase, relative interferometric phase is also wrapped. Although the notation and the equations are rather similar, the relative interferometric phase is the difference between two pixels and the range difference Δr is the range difference between the two pixels (Figure 3.5).

For large distances r and short baselines $B_\perp$, as we have in SAR remote sensing, the range difference Δr that is caused by the height difference Δh can be simplified as

$$
\Delta r \approx \frac{B_\perp}{r \cdot \sin\theta}\Delta h
\tag{3.9}
$$

Therefore,

$$\Delta\phi = \frac{4\pi}{\lambda} \frac{B_\perp}{r \cdot \sin\theta} \Delta h \qquad (3.10)$$

3.2   SAR Interferometry Processing Steps

SAR interferometry requires a series of processing steps. In this chapter, we follow the processing step by step, explaining the necessity for each of the steps and possible problems along the way.

3.2.1   SAR Data Selection: Considerations for SAR Interferometry

The very first step is the selection of the data. This is certainly application-dependent, but also often a matter of data availability. Although we should select the right data for the task, often we will end up having to use the available data. As a rule of thumb, shorter wavelength data, like X- or C-band, can provide higher precision in interferometric processing, while longer wavelengths, like L-band, are less affected by temporal decorrelation. We may therefore prefer X- or C-band for applications in urban areas, where we have significantly less temporal decorrelation, while L-band is more suited for applications in vegetated areas.

We prefer short temporal baselines, so that the temporal decorrelation is reduced. For the perpendicular baseline, this is a trade-off between the higher theoretical precision we can achieve with large baseline data against the larger problems such data can cause in the phase unwrapping process. We will discuss this in more detail later on.

For interferometric processing, we need amplitude and phase information. The data therefore have to be available in the so-called single-look complex (SLC) format. If the data are delivered in raw format, the SAR image must first be focused to have an SLC image appropriate for interferometric processing.

3.2.2   SAR Co-Registration

The first step is the co-registration of the SAR data to a common radar coordinate system. The so-called slave image is co-registered into the coordinate system of the so-called master image. Precise co-registration makes sure that the scatterers in both images are identical, as mentioned, this is a condition for SAR interferometry to yield meaningful results. In general, a co-registration precision of 1/10th of a pixel is required. Some acquisition modes require even higher co-registration precision for SAR interferometry. SAR co-registration is the first step in interferometric SAR processing. It is a standard procedure, and in general, the accuracy requirement of 1/10th of a pixel is achieved with the methods described in the following. However, co-registration can also fail, for example, when large changes on the ground hinder the successful identification of homologous points in both images. It is therefore mandatory to carefully analyze the co-registration results for possible co-registration errors, as errors at this early stage could render the complete process useless.

3.2.3   Coarse-to-Fine Co-Registration

The standard procedure for SAR image co-registration follows a coarse-to-fine strategy, where the slave image is first roughly, i.e. within a few pixels, co-registered to the master image. Afterwards, this co-registration is refined and the resampling parameters are calculated. Finally, the slave image is resampled. The processing steps and their detailed implementation can differ in the details.

3.2.3.1   Coarse co-registration

The goal of the coarse co-registration is to identify homologous pixels in the master and slave images and to estimate the coarse shift in pixels between the master and slave images. A precision within a few pixels is required. Using modern sensor systems, coarse co-registration can rely solely on the orbit information delivered with the data, as this is often precise enough to achieve a relative precision of a few pixels.

Otherwise, a few search points are distributed over the image. Within a search area, the pixel combination with the highest normalized correlation σ is selected, where σ is calculated within a search window of N pixels, with $m_{x,y}$ and $s_{x,y}$ representing the pixel values at x, y in the search window of the master and the slave, respectively. $\bar{m}$ and $\bar{s}$ are the mean values and σ_m and σ_s are the standard deviations of the values in the search window of the master and slave images:

$$\sigma = \frac{1}{N-1} \sum_{x,y} \frac{(m_{x,y} - \bar{m})(s_{x,y} - \bar{s})}{\sigma_m \sigma_s} \tag{3.11}$$

The average or median of the best pixel combinations is then used as the estimated shift between slave and master.

Nowadays, with many SAR satellite systems providing very precise orbit information, the coarse co-registration can often be replaced with an estimation of the coarse offset based solely on the given orbits.

3.2.3.2   Fine co-registration

Under fine co-registration, we understand the process of finding homologous sub-pixels in the master and slave images and establishing co-registration pairs within sub-pixel precision.

The coarse co-registration step is necessary to drastically reduce the search area beforehand, thus allowing for sub-pixel precision in the fine co-registration step. To achieve sub-pixel precision, the images or the image areas are oversampled. Afterwards, pixels are matched. In contrast to the coarse co-registration, this is often done in the frequency domain after a Fast Fourier Transformation. However, sub-pixel matching of strong scatterers can also be applied in the spatial domain.

Fine co-registration is performed on a relatively large number of points (often 1000 or more), so that a large overestimation of the linear equations in the following step is possible. This will allow the removal of a large number of outliers, stabilizing the process and thus meeting the required sub-pixel accuracy of 1/10th of a pixel.

3.2.3.3   *Co-registration parameter estimation*

Based on the previous fine co-registration results, a set of linear equations is set up to estimate the co-registration parameters within an overestimated system of linear equations.

It is common to assume a polynomial transformation between the slave and master images. The question is the degree of the polynomial. Assuming similar resolution, an affine transformation would be suitable. However, it is also not uncommon to estimate a second-degree polynomial. This allows for a better fitting result when there are differences in the acquisition parameters, but should be taken with care as it also allows for a greater degree of freedom and the introduction of more subtle and harder-to-recognize co-registration errors.

The process is normally implemented iteratively, estimating the parameters based on all point pairs in the first step, calculating the residuals for each pair, removing outlier pairs, and re-estimating the parameters with the remaining pairs. This is repeated until the number of remaining pairs is below a threshold or the residuals are all below a certain threshold, so the resampling parameters can be estimated.

3.2.3.4   *Slave image resampling*

Finally, the slave image is resampled into the master image coordinate system. First, an empty image in the master image coordinate system is created. For each pixel in the newly formed image, the corresponding pixel coordinates in the slave image are calculated based on the resampling parameters estimated in the previous steps. The pixel value is calculated based on the resampling kernel, which can be as simple as a bilinear transformation and as complex as a cosine function kernel. This process is comparatively time-consuming. The product of the resampling process is a resampled slave image in the coordinate system of the master image. Therefore, each pixel in the master image corresponds to the exact same pixel in the resampled slave image.

3.2.4   Orbit and DEM-based Co-Registration

Another approach relies on the precise orbit information of modern SAR satellite systems and the availability of a DEM. The precision requirements for the DEM are rather lax, as a height error still leads only to a small shift between the master and slave images in an interferometric pair with baselines below the critical baseline. See the following for more information on the critical baseline. However, the orbit information needs to be precise. Modern satellites provide very precise orbit information, but sometimes it is necessary to wait several days after image acquisition to get the precise orbit ephemerides, as they are processed after the fact based on measured satellite positions.

In this process, a subset of the DEM is generated that covers the footprint of the image. For each pixel of the DEM, the corresponding master image coordinates and slave image coordinates are calculated. Afterwards, the data are triangulated based on the master image coordinates, with the height and slave image coordinates as information for each point in the triangles.

Now, for each pixel in the master image, the corresponding triangle is searched and the height and slave coordinates are interpolated based on the triangle points, so that for each pixel in the master image, a corresponding height value and slave coordinate are received. Finally, the value of the resampled slave image is interpolated from the original slave image based on the derived slave image coordinates.

This process allows for precise geocoding and works well even for images with large changes or with a limited number of identifiable homologous points, depending on the availability of precise orbit information and a good DEM. The precision requirements for the orbit and the DEM are higher for high-resolution images and comparably low for low-resolution images.

It is generally possible to achieve a precision of 1/10th of a pixel with this method alone, but often an additional post-processing step is required to further enhance the precision of the orbit.

Enhanced spectral diversity for TOPS

In the Terrain Observation with Progressive Scanning (TOPS) mode, a fast steering of the antenna beam in the azimuth direction is used as a correction for the scalloping effect that can appear in ScanSAR images. Due to the speed of antenna beam steering however, targets at different azimuth positions are acquired at different squint angles, which results in variations in the Doppler centroid within a burst (Rodriguez-Cassola *et al.*, 2015). The large variation in the Doppler centroid can lead to many artifacts in SAR interferometry, requiring a correction of the Doppler centroid shift similar to the operation used in Spotlight interferometry. However, in TOPS, a more accurate estimation of the Doppler centroid is necessary, which requires a more precise resampling. In TOPS, phase discontinuities may appear in overlapping regions between consecutive bursts due to differences in the Doppler centroid (Prats-Iraola *et al.*, 2012). In order to avoid phase discontinuities in these areas, we have to keep phase error below 3°. This means that we have to reach very high co-registration accuracy compared to the accuracy required in the stripmap acquisition mode, where the squint angle is fixed over the whole acquisition time. For Sentinel-1A data acquired using the Interferometry Wide (IW) mode, where the variation of Doppler centroid is about 5.5 kHz, a co-registration accuracy of around 0.0009 of a pixel is required in order to get a usable interferogram, i.e. without phase discontinuities (Yague-Martinez *et al.*, 2016). The standard co-registration model as described here does not provide such a high accuracy, it generally offers 0.1–0.01 co-registration accuracy, which is not enough in the case of TOPS. In order to remove this source of error, we can, for example, use the enhanced spectral diversity (ESD) method (Prats-Iraola *et al.*, 2012). ESD is an improvement to the spectral-diversity (SD) method, presented by Scheiber and Moreira

(2000). In ESD, we use the overlapping regions in the TOPS mode acquisition as consecutive bursts have a large spectral separation in these overlapping areas. Therefore, the basic idea to enhance SD is to apply the same procedural steps to these overlapping regions.

In ESD, each of the two corresponding bursts (master and corresponding slave bursts) is divided into two parts, also called two looks, where each look has a different center frequency. Then, the offset between the master and slave bursts is computed directly from the differential interferogram between two low resolution interferograms obtained from the master and slave parts. The final offset is computed by averaging all values obtained for each set of corresponding master and slave bursts.

The accuracy of the ESD can be calculated via (Prats-Iraola *et al.*, 2012)

$$\sigma_{ESD} = \frac{\sqrt{2} \cdot \sigma_{\Theta}}{2\pi \cdot \Delta f_{dc} \cdot c \cdot \tau} \qquad (3.12)$$

where τ is the imaging sampling in azimuth, Δf_{dc} is the difference in the Doppler frequency and σ_{Θ} is the standard deviation of the phase noise, which can be approximated as

$$\sigma_{\Theta} = \frac{1}{\sqrt{2N}} \frac{\sqrt{1 - \gamma^2}}{\gamma} \qquad (3.13)$$

where N is the number of averaged samples and γ is the coherence value of the interferogram (see the following for an explanation of coherence). Thus, by increasing the number of N, the accuracy can be improved. ESD is the standard, but not the only method, for the co-registration of Sentinel-1 images. It is preceded by a coarse co-registration to the pixel level using a DEM-based co-registration approach using the provided orbit information. The precise orbit ephemerides are strongly recommended for use in this operation to ensure high precision of the first co-registration step before applying ESD. Generally, DEM-based co-registration with ESD is a stable and in most cases a suitable co-registration strategy. Problems may occur if wide area motion is disturbing the phase information ESD is using for co-registration.

3.2.5 Interferogram Generation

After the co-registration of the master and slave images, the generation of the interferogram is rather trivial. The interferogram is defined as $\phi_1 - \phi_2$, or the phase of image 1 minus the phase of image 2. As we are dealing with complex data, this is typically implemented as a multiplication of the complex conjugate:

$$v_{i,j} = u_{1i,j} \times u^*_{2i,j} \tag{3.14}$$

Multiplying with the complex conjugate results in a difference between the phases and a multiplication of the amplitude. Be aware that it is also possible to implement just the phase difference, but this would require to call a subsequent wrapping function to keep the phase wrapped.

The resulting interferogram (Figure 3.6) represents phase, or distance changes as a color map in which we find that we have the same color again if the distance changes with

$$\Delta r = \frac{\lambda}{2} \tag{3.15}$$

In the interferogram, these phase changes are dominated by changes in the range direction, as the distance changes with an increasing distance between the sensor in the range direction. This is the so-called flat-Earth phase component that we want to remove in the next step.

Figure 3.6　Interferogram of Uluru/Ayers Rock.

3.2.6　Flat-Earth Phase Removal

In this step, we remove this phase. The phase at a height of 0 m is simulated at several positions in the image, which is also why this processing step is called the flat-Earth phase removal. Afterwards, the phases between these points are interpolated. Finally, we remove the simulated phase from the interferogram and receive a so-called flat interferogram (Figure 3.7).

This already includes more information and we can see now the influence from the local topography in the flat interferogram.

As before, for two pixels with the same range, the height difference Δh depends on

$$\frac{\Delta \phi}{\Delta h} = \frac{4\pi}{\lambda} \frac{B_\perp}{r \cdot \sin \theta} \tag{3.16}$$

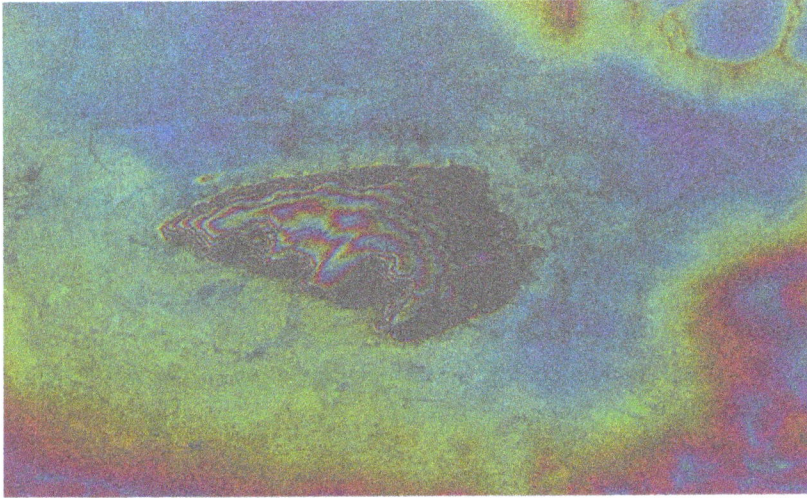

Figure 3.7 Interferogram of Uluru/Ayers Rock after flat-Earth removal.

3.2. Topographic phase:

$$\Delta\phi = \frac{4\pi\Delta h}{\lambda}\frac{B_\perp}{r\cdot\sin\theta} \tag{3.17}$$

where $\Delta\phi$ is the phase difference between the points and θ is the incidence angle. After removing the flat-Earth phase, this process can also be defined as a removal of the range difference between the points, so that the equation as shown above is true between all pixels, as the condition of having the same range now applies.

3.2.7 Height of Ambiguity

Based on this equation, we can calculate the height of ambiguity. Since the phases are wrapped between $-\pi$ and π, only height differences between the 2π range are unambiguous. Other heights are ambiguous because of the phase wrapping. To calculate the height of ambiguity, we define $\Delta\phi = 2\pi$ and get

$$\Delta h = \frac{\lambda}{2B_\perp} \cdot r \cdot \sin\theta \qquad (3.18)$$

From this, we can infer that the larger the perpendicular baseline $B_\perp$, the smaller the height of ambiguity. This also means that higher height accuracy is achievable.

Example 3.1 If we assume two image pairs acquired from a C-band sensor at 45° incidence angle, one pair has a perpendicular baseline $B_{\perp,1} = 250$ m and the second one $B_{\perp,2} = 50$ m.

$$\Delta h_1 = \frac{0.0515 \text{ m}}{2 \times 250 \text{ m}} \times 500 \text{ km} \times \sin 45° = 36 \text{ m}$$
$$\Delta h_2 = \frac{0.0515 \text{ m}}{2 \times 50 \text{ m}} \times 500 \text{ km} \times \sin 45° = 182 \text{ m} \qquad \square$$

The height of ambiguity changes from 36 m to 182 m with the change in the perpendicular baseline. Larger baselines mean higher sensitivity to height.

That is to say, for an interferogram of the first image pair, the height difference between lines of the same color is 36 m, while in the second image, that would be 182 m. But, as the heights are ambiguous, in the first interferogram, points with the same phase value might have a height difference of 0 m, 36 m, −36 m, 72 m, −72 m, 108 m, etc. This ambiguity in height is due to the wrapped phase and has to be solved before creating a DSM. This is done via phase unwrapping, a process described in the following.

However, with the dense fringes—fringes are the lines of the same color in the interferogram—phase unwrapping is considerably more difficult and erroneous in the first image with a height of ambiguity of 36 m than in the second image with 182 m height of ambiguity.

Meanwhile, a small height of ambiguity also allows for a more precise height measurement. We express errors in an interferogram, e.g. caused by noise, in terms of degrees of the phase. So, assuming that we can measure within 5° of the phase, then, taking the equation for the calculation of the topographic height difference between two points from the phase difference

$$\frac{\Delta\phi}{\Delta h} = \frac{4\pi}{\lambda}\frac{B_\perp}{r \cdot \sin\theta} \qquad (3.19)$$

we can form

3.3. Height from interferometric phase:

$$\Delta h = \frac{\lambda\Delta\phi}{4\pi}\frac{r \cdot \sin\theta}{B_\perp} \qquad (3.20)$$

---

so that the following are obtained.

Example 3.2

$$\Delta h_1 = \frac{0.0515 \text{ m} \times 0.0873}{4\pi} \times \frac{500\,000 \text{ m} \times \sin 45°}{250 \text{ m}} = 0.5 \text{ m}$$

$$\Delta h_2 = \frac{0.0515 \text{ m} \times 0.0873}{4\pi} \times \frac{500\,000 \text{ m} \times \sin 45°}{50 \text{ m}} = 2.5 \text{ m} \qquad \square$$

Thus, assuming that we can measure within 5°, this calculation leads to a precision of 0.5 m in the first image pair with a 250 m perpendicular baseline, but only to 2.5 m within the second image pair with the 50 m baseline. A smaller height of ambiguity helps us obtain DSMs with theoretically higher precision, but with an increased difficulty during phase unwrapping and a higher sensitivity to noise, which will be discussed in more detail in the following.

In Figure 3.8, this is demonstrated with three interferogram examples over Las Vegas. The interferogram in Figure 3.8(b) has a perpendicular baseline of only 7.8 m. So, the influence of the topography on the phase is very limited and we do not find dense fringes along the building facades. In Figure 3.8(c) the perpendicular baseline is 110 m, and in Figure 3.8(d) the perpendicular baseline is −245 m. Accordingly, the fringe density increases along the facades, clearly demonstrating the

relation between the perpendicular baseline and the fringe density and
therefore the height of ambiguity.

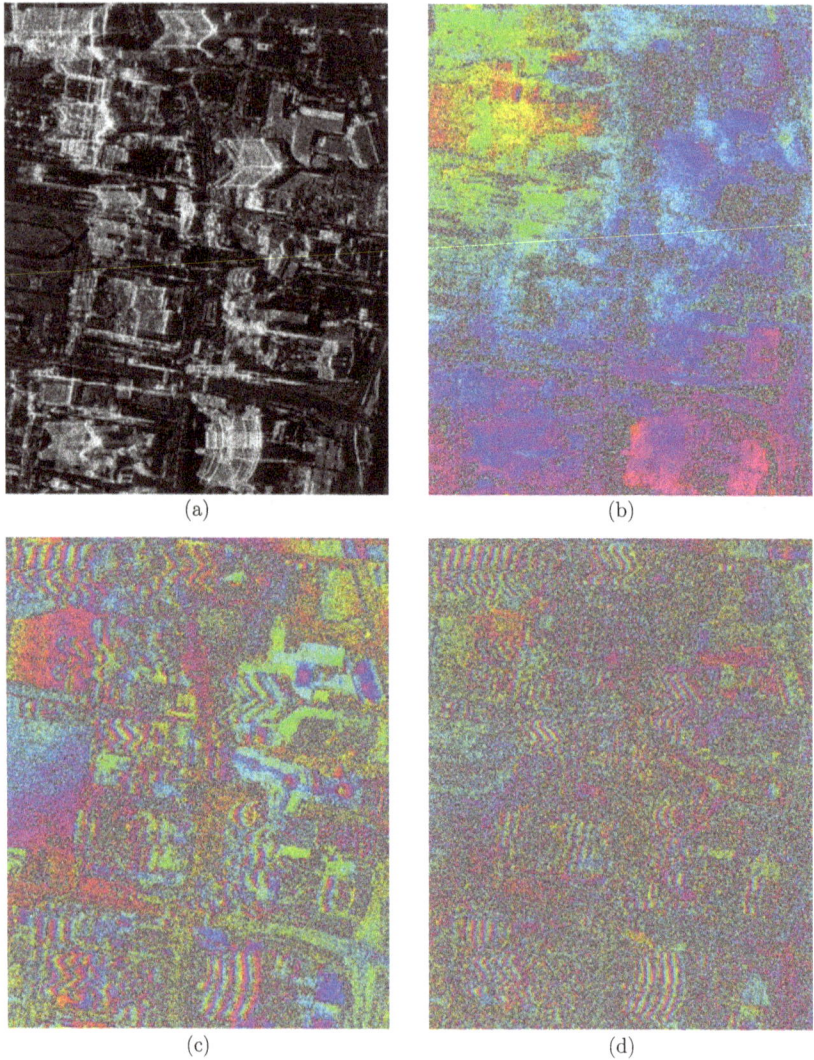

Figure 3.8   (a) Master TerraSAR-X high-resolution spotlight image over Las Vegas
from 2019-09-08 (© DLR, 2010); (b) Interferogram master with image from 2010-06-01
with 7.8 m baseline; (c) Interferogram master with image from 2010-12-16 with 110 m
baseline; (d) Interferogram master with image from 2010-02-22 with −245 m baseline.

3.2.8   Topographic Phase Removal

The next step is not mandatory for DSM generation, but can be useful. Similar to the flat-Earth removal—in fact so similar that both steps can be combined in a single process—known phase contributions from a previously existing DEM can be removed from the data. This can be helpful during further processing. After the removal of the topographic phases from a previously existing DEM, the resulting heights will be relative to this pre-existing DEM, so that a height difference between two pixels of 5 m would mean that there is a height difference between the DEM height of pixel 1 and the DEM height of pixel 2 of 5 m. The phase for each point is calculated based on

$$\phi = \frac{4\pi h}{\lambda} \frac{B_\perp}{r \cdot \sin\theta} \tag{3.21}$$

where h is the height above the estimated zero height from the flat-Earth removal. The calculated phase ϕ is then removed from the flat-Earth interferogram.

For this process to work, h needs to be known for each pixel in the interferogram. This requires a transformation of the height model into the radar coordinate system. It is theoretically possible to transform each pixel during the process in world coordinates, but this process requires knowledge of the height and could therefore only be implemented indirectly with an iterative process, which does make it significantly slower than transforming the DEM into radar coordinates.

The transformation of the DEM into radar coordinates is similar to the process described in Section 3.2.4 of Chapter 3. Each point in the DEM is transformed in radar coordinates, which is an unambiguous transformation as the height and the position in the world coordinates are known. Afterwards, the transformed coordinates are triangulated, for example, using a Delaunay triangulation. Then, for each pixel in the radar coordinate system, a matching triangle is found, and the height is interpolated based on the heights of the triangle points. Other interpolation methods are also usable and may provide even better results.

Now, we can calculate the phase for each pixel in the radar coordinate system. However, as this may take some time, this is typically avoided by calculating the phase for each point in the DEM and transforming the height and the phase to radar coordinates in a single step and then triangulating those points. Now, instead of finding the height for each point and processing the phase, we can directly interpolate the topographic phase value for each pixel based on the topographic phase value of each triangle edge. This can save significant processing time and is therefore the typical implementation. Figures 3.9 and 3.10 show examples of DEM converted to radar coordinate system and interferograms after removal of phase, respectively.

3.2.9   Coherence Estimation

An important indicator for the expected quality of an interferogram is the coherence. We define two phases as coherent when they have a

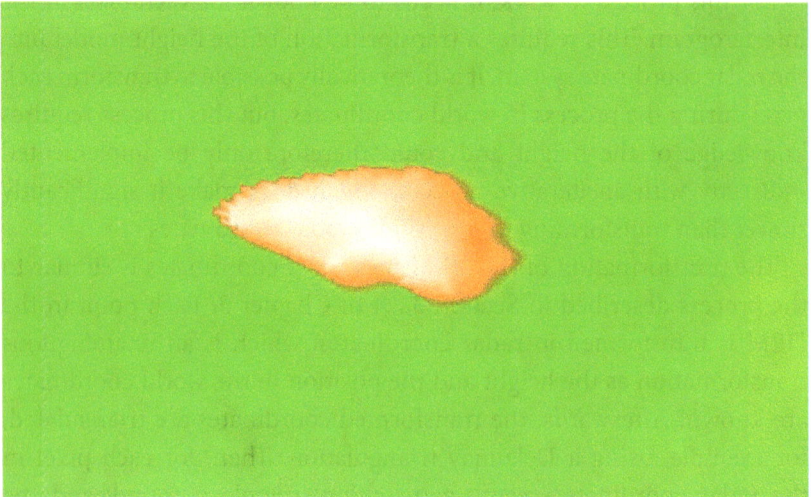

Figure 3.9   Interpolated height for the phase calculation for the InSAR pair over Uluru/ Ayers Rock.

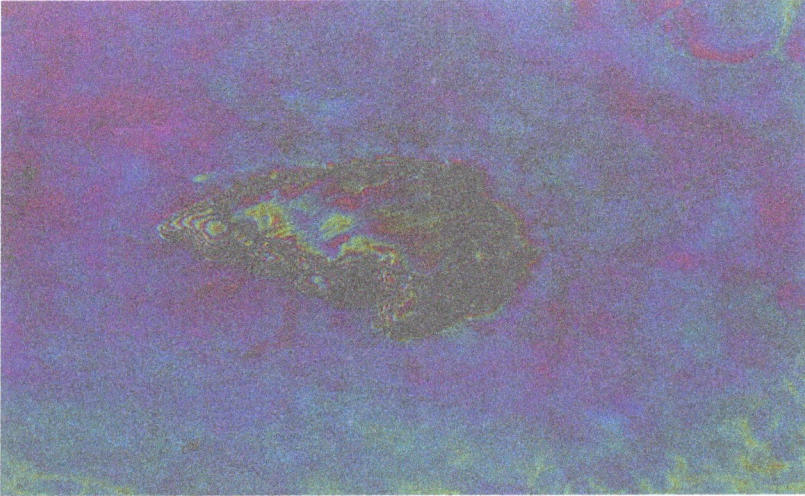

Figure 3.10 Interferogram of Uluru/Ayers Rock after removal of the topographic phase.

constant phase difference, which is a prerequisite for working on phase interference and creating (meaningful) interferograms. Furthermore, the phases need to be at the same frequency and waveform to be considered coherent, which is normally the case in SAR interferometry.

Given two SAR images, the complex coherence γ between the two images can be calculated:

3.4. Spatial coherence:

$$\gamma = \frac{E[v_1 v_2^*]}{\sqrt{E[v_1 v_1^*]}\sqrt{E[v_2 v_2^*]}} \tag{3.22}$$

where v_1 is the complex value of the first image and v_2 is the complex value of the second image. $v_{1,2}^*$ is the complex conjugate and $E[...]$ is the expected value. γ is between $0 < \gamma < 1$.

Due to the random nature of the speckling effect, we have to deal with expected values. Therefore, coherence is estimated based on several pixels, typically in the form of a box window around the pixel to be

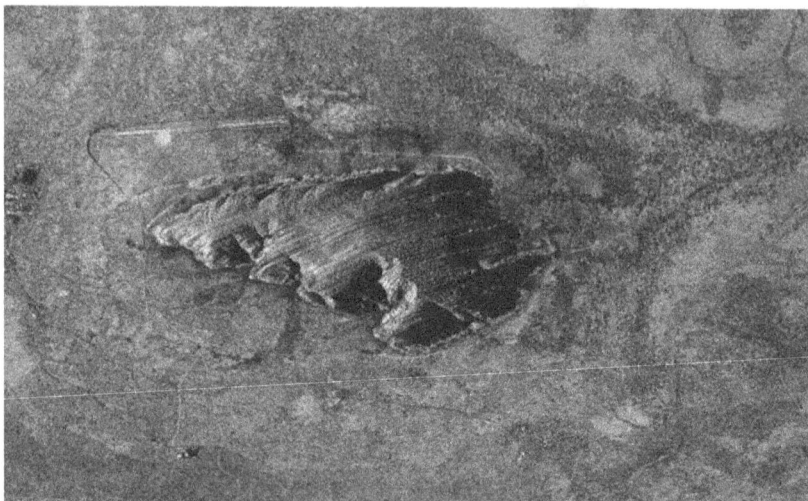

Figure 3.11   Coherence image of the interferometric pair over Uluru/Ayers Rock.

analyzed and the expected value $E[...]$ is the aggregated mean value in the box.

Coherence images are very useful to estimate the quality of interferograms extracted from different parts of the images. There are various reasons for the loss of coherence in an image. They will be discussed in more detail in Section 3.4 of Chapter 3. Such a quality indicator is extremely useful in the processing, as it allows us to estimate the expected quality of our results in the early stages of processing. It furthermore gives us a spatial indication of areas with acceptable and unacceptable coherence, allowing us to mask out areas of low coherence from further processing. Figure 3.11 shows an example of coherence map.

This is especially important for phase unwrapping, where unwrapping extremely noisy areas, i.e. areas with a low coherence, can lead to many phase unwrapping errors and can also be extremely time-consuming, which is unnecessary as the results are anyway dominated by noise in such areas. It is therefore quite common to estimate the coherence and mask out areas of low coherence in the phase unwrapping step.

3.2.10 Phase Unwrapping

In the phase unwrapping step, the wrapped phases are to be unwrapped, so that the limitations from staying within $-\pi$ to π are resolved. Our interferometric phase consists of

$$\phi = \phi_t + \phi_n$$

where ϕ_n is the phase noise and ϕ_t contains the true phase information, like, for example, the topography-induced phase. However, in our SAR images, we deal with the wrapped phase φ, which is

$$\varphi = W\{\phi\}, \quad -\pi < \phi < \pi \qquad (3.23)$$

where $W\{\cdots\}$ is the wrapping operator. The unwrapping problem is to find from a given wrapped phase φ the estimation of the unwrapped phase $\hat{\phi}$ or, ideally, of $\hat{\phi}_t$. However, ambiguity is the main problem in the phase unwrapping process. There are several solutions available; during the phase unwrapping process, we try to find the most likely phase. This phrasing for me, and I guess for some readers, does not fan the flame of confidence in the process.

Due to its ambiguous nature, additional constraints are required in the phase unwrapping process. Such constraints can be a 'smoothness' requirement, where we expect the phase to change smoothly, which can be a useful constraint in the unwrapping of topographic phases of natural objects like mountains. In urban areas, however, a smoothness requirement may not lead to ideal results and we can include other constraints, e.g. building footprints, to unwrap the phases along building facades.

However, there is a clear relationship between φ and ϕ as

$$\phi = a \cdot 2\pi + \varphi \qquad (3.24)$$

where a is an integer value. The fact that the unknown a is an integer helps in reducing the search space and offers means to solve the unwrapping problem.

3.2.10.1   1D phase unwrapping

Looking at phase unwrapping in only one dimension (Figure 3.12) means that any phase discontinuity will cause a disconnection. Assumptions on phase discontinuities need to be based on a model of the phase behavior along one dimension to be successful. For example, phase values that change linearly can be successfully unwrapped in one dimension as long as that linearity is known and can be well estimated. More complex or unpredictable phase changes on the other hand cannot be recovered from only one dimension.

3.2.10.2   2D phase unwrapping

Two-dimensional phase unwrapping offers more possibilities to find paths along which the phases can be unwrapped, improving the connectivity of the phases. Furthermore, phase discontinuities can be detected in 2D phase unwrapping from residues in the data. A residue is a point around which the integration of the phase gradient does not return zero (Goldstein *et al.*, 1988). Finding such residues is, therefore, the key to detection and correction of such phase discontinuities.

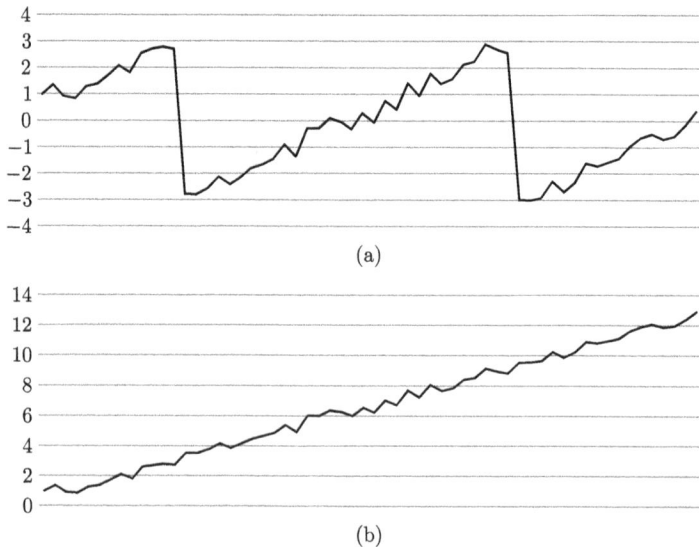

(a)

(b)

Figure 3.12   1D wrapped (a) and unwrapped (b) phase series.

3.2.10.3 Phase unwrapping summary

Phase unwrapping is the process of recovering the unambiguous phase values from the known ambiguous wrapped phase values between $-\pi$ and π. It is an important process in InSAR processing. For phase unwrapping in InSAR, reliable 2D phase unwrapping algorithms are needed and have now been developed for decades (Chen & Zebker, 2001; Costantini, 1998; Ghiglia & Romero, 1996; Goldstein *et al.*, 1988).

3.2.11 DSM Generation

3.2.11.1 Phase-to-height conversion

After the unwrapping, the unwrapped phase is converted to height values using

$$\Delta h = \frac{\Delta \phi \cdot \lambda \cdot r \cdot \sin \theta}{4 \pi B_\perp} \tag{3.25}$$

As we calculate Δh, which is the height difference between two pixels, it becomes clear that we are dealing with relative height values and not absolute values. Since even after phase unwrapping, we only know the relative unwrapped phases between pixels and not the absolute number of phase cycles that passed.

These relative heights can then be formed into a DSM, as shown in Figure 3.13. However, as relative values, such a DSM is of only limited value. A correction for absolute height values is therefore necessary.

3.2.11.2 From relative to absolute height values

To get from the relative Δh to an absolute h value, additional information is needed. This can come in the form of a reference point with a known height. Just one reference point is needed as we know the relative height difference between all the pixels in the image, with the exception of pixels we masked out due to low coherence during a previous step.

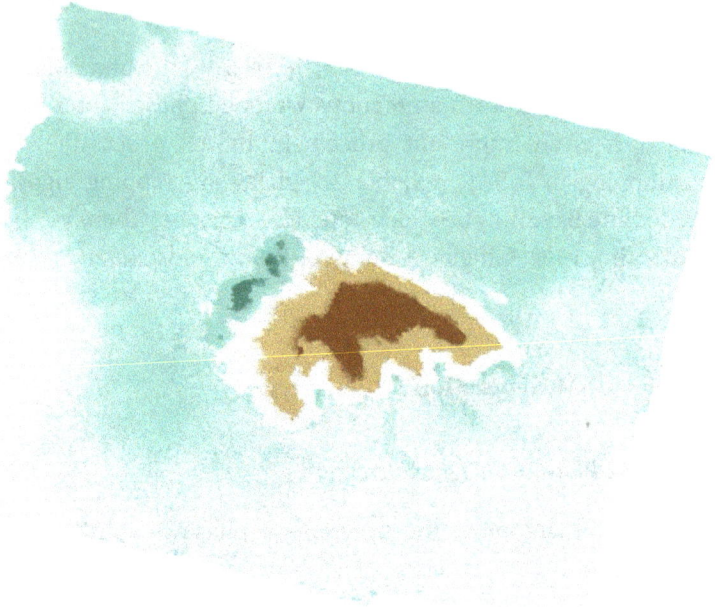

Figure 3.13 Digital surface model of Uluru/Ayers Rock derived from interferometric SAR.

In this regard, if large areas have been masked and we find separated image parts, or islands, the estimation of the unwrapped phase will be separated for these parts and phase unwrapping will not provide unified results. In such cases, more than one reference point might be necessary, so that for each part of the unwrapped image, or for each "island", there is an available reference point.

Instead of well-known reference points, a pre-existing DEM can be used. In this case, the average height of the relative DSM can be matched to the average DEM height to get a DSM in absolute values.

3.2.11.3 Georeferencing of the final DSM

The final step is the georeferencing of the DSM, which is in radar coordinate systems and should be referenced to a world coordinate system or a local coordinate system. SAR radar coding, i.e. the transformation of 3D world coordinates into 2D image coordinates, is well defined via the Range-Doppler process (Cumming & Wong, 2005). However, the transformation from 2D image coordinates into 3D world coordinates is ambiguous.

Knowing the height of the point in world coordinates though allows for an unambiguous transformation from radar image coordinates and height into 2D world coordinates. As a result of the process, we get a point cloud of 3D coordinates because the resulting points in world coordinates will not be in a regular raster due to the side-looking and run-time geometry of SAR images.

In the final step, the point cloud data needs to be interpolated into a regular raster. There are several ways to do it, a standard approach would be to triangulate the points and then interpolate the heights. This approach works well in natural environments, where we can assume a 2.5D topography, i.e. a topography without overhanging walls. In cities with large skyscrapers, this approach often delivers unsatisfactory results. An interpolation that is aware of building structures could be more suitable.

3.3 SAR Interferometry with Spotlight and TOPS Images

The previous examples have all been in the standard operating mode of the SAR system, the so-called stripmap mode. However, advanced SAR modes, like the spotlight mode or the TOPS mode, require some additional steps for InSAR processing.

In the spotlight mode, the synthetic antenna is enlarged by sweeping the azimuth beam during image acquisition, thus increasing the illumination time t_{AP}. According to Eineder *et al.* (2009), the illumination time can be expressed by

$$t_{AP} = \frac{B_{DP}}{\dot{f}_{DC} - FM} \tag{3.26}$$

where B_{DP} is the antenna Doppler bandwidth, $\dot{f}_{DC}$ is the Doppler rate caused by beam sweeping, and FM is the frequency modulation rate. Therefore, the effective azimuth time interval of the focused spotlight image Δt_{SSC} is shorter than the raw data time interval Δt_{raw}.

$$\Delta t_{SSC} = \Delta t_{raw} + \frac{B_{DP} - \Delta t_{raw}\dot{f}_{DC}}{FM} \qquad (3.27)$$

In SAR spotlight mode images, there is a systematic Doppler centroid shift in azimuth direction, which needs to be considered in InSAR processing as well as during filtering steps. Therefore, the shift in the Doppler needs to be accounted for during resampling as well as for filtering operations like azimuth bandwidth filtering.

If these shifts are not accounted for correctly, phase residuals in azimuth are to be found, as demonstrated in Figure 3.14, where such errors are clearly visible in the top and bottom of the interferogram.

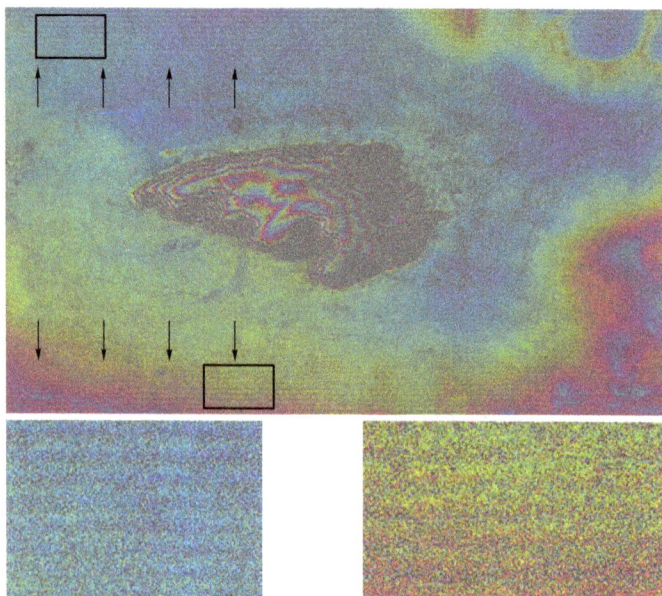

Figure 3.14 Interferogram of Uluru/Ayers Rock after flat-Earth removal but without correction for the Doppler shift.

Similarly, the Doppler centroid shift must also be corrected in TOPS mode data (Prats-Iraola *et al.*, 2012). Additionally, TOPS mode data need to be resampled with a very high precision. This can be achieved with SD or in the TOPS case also with ESD, as discussed in the previous section on ESD.

3.4 Error Sources

In theory, we can achieve precise DSM from SAR interferometry. In reality, however, various error sources disrupt the results. It is therefore very important to know the most common error sources and their properties.

3.4.1 Loss of Coherence

The loss of coherence is a symptom rather than the problem. Thanks to the possibility of estimating the coherence—remember we are not actually measuring the coherence, but rather estimating it based on the similarity between the images—we have a high-quality indicator for interferometry.

The loss of coherence is not actually an error source. Some errors lead to a loss of coherence, but there are several error sources in interferometry. So, if an error is described as loss of coherence without further explanation, it is safe to assume, and not uncommon, that the actual error source is unknown.

There are many possible factors influencing coherence, and therefore the interferometric quality, so it is seldom possible to clearly single them out. In such cases, the error is also often just described as loss of coherence, in the same way as we sometimes describe illnesses as "having a fever", which is also only a symptom, while not knowing the true reason. As for common cold, it is sometimes not necessary to know the exact reason to get better, so it is often not necessary to know the exact reason for the loss of coherence, although, similar to cold, it can help in mitigating the effects.

Coherence can also be used as a tool for classification. The relation between the temporal baseline, wavelength, and loss of coherence can help in estimating the amount of changes on the ground, assuming that the loss of coherence is mostly attributed to temporal decorrelation. This can be useful for classification and mapping purposes.

SAR coherence has been used in various applications over the years. Prati and Rocca (1992) produced coherence maps for target classification, while Bruzzone *et al.* (2004) proposed a novel system for the classification of SAR images based on concepts of long-term coherence and temporal backscattering variability. Additionally, one well-established application of SAR is the detection of temporal changes in a scene through the coherence change detection (CCD) (Preiss & Stacy, 2006). The coherence and intensity characteristics of SAR images have been exploited in techniques to monitor urban activities and changes. Unsupervised thresholding techniques for change detection using the coherence and intensity characteristics of SAR imagery have been proposed in a number of studies (He & He, 2009). Jendryke *et al.* (2017) combined social media messages with SAR images to express human activities and urban changes in Shanghai. Washaya *et al.* (2018) used CCD for the damage assessment of various natural and man-made disasters.

3.4.2   Temporal Decorrelation

Changes on the ground occurring between acquisitions can lead to temporal decorrelation. The changes are related to physical changes on the ground affecting the distance between a sensor and the backscattering object. Changes that exceed half a wavelength lead to a total decorrelation due to the wrapping of the phase. Smaller changes also lead to a reduction in the coherence of the phase.

Buildings, for example, typically do not change much and can keep their position and physical appearance over long periods, staying coherent. Trees, on the other hand, and especially their leaves and twigs, are moving constantly. They are growing and shaking with the wind. In the several days of time difference images between repeat-pass satellite SAR acquisitions, the physical appearance of a tree can change significantly between the moments of image acquisitions.

How changes effect the coherence can be estimated, for example, with a rather simple model (Zebker & Villasenor, 1992):

$$\gamma = e^{-\frac{1}{2}\left(\frac{4\pi}{\lambda}\right)^2 (\sigma_y^2 \sin^2 \theta + \sigma_z^2 \cos^2 \theta)} \tag{3.28}$$

where σ_y and σ_z describe the motion of the object, as shown in Figure 3.15.

As we can see from Figure 3.16, the loss of coherence depends strongly on the wavelength of the system. For an X-band system with about 3 cm wavelength, even a slight motion can quickly diminish the coherence below the 0.3 value that is considered still usable. Furthermore, smaller wavelengths like X-band are mainly reflected from the top of the canopy; they are sensitive to leaves and smaller twigs, which change the fastest. Longer wavelengths, like, e.g., L-band with about 20 cm wavelength, are backscattered only by larger elements and thicker twigs or the stem of the tree. These move less in the wind and grow slower. Therefore, they

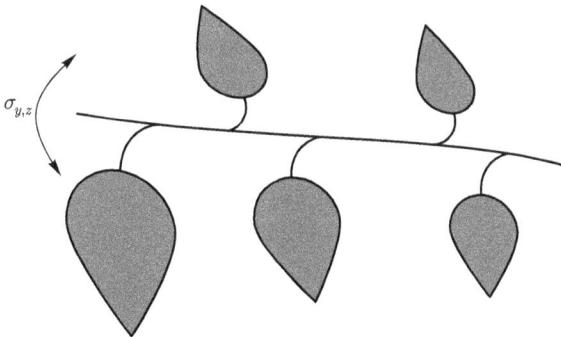

Figure 3.15 Motion of leaves along $\sigma_{y,z}$.

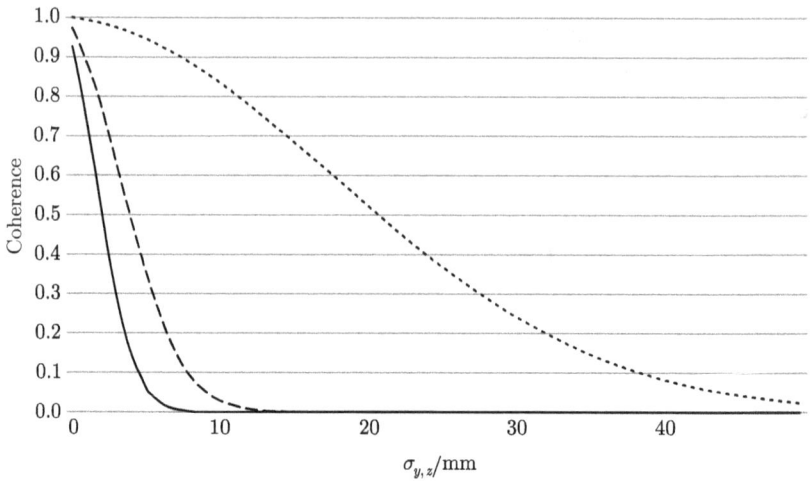

Figure 3.16 Loss of coherence following Zebker and Villasenor (1992). X-axis is the motion of $\sigma_{y,z}$ in mm. Solid line: X-band; dashed line: C-band; dotted line: L-band.

are much less sensitive. This is the reason why long wavelength systems can keep coherent over vegetated areas; but, they are also not immune against temporal decorrelation.

3.4.3 Atmospheric Delay

In many SAR algorithms, we assume that the signal travels at the speed of light. This however is a simplification, as an electromagnetic signal slows down on its way through the atmosphere. In interferometric SAR, this general slowing down is not of great importance because we work with phase differences between pixels. That is to say, as long as the atmospheric delay is identical between the pixels, or more precisely, as long as the difference between the atmospheric delays of the two images forming an interferogram is identical between the pixels, the calculated phase difference is not affected and interferometric processes work just fine. The problem in terms of atmospheric delay in InSAR is caused by the differences in the atmospheric delay over an image. Such differences cause phase differences from one pixel to another pixel, including a phase component from the difference in the atmospheric delay. What causes this difference in the atmospheric delay?

On the way from the satellite sensor to the ground, the signal is going through the ionosphere. In the ionosphere, the path delay depends on the number of electrons in the ionosphere TEC:

$$\delta_{\text{ionosphere}} = \frac{40.28 \ \text{m}^2 \cdot \text{s}^{-2}}{f^2} \frac{\text{TEC}}{\cos \theta} \tag{3.29}$$

where f is the center frequency of the radar and TEC is the number of electrons given in electrons per square meter, where typical values are 5–10 TECU, where one TECU is 10^{16} electrons. For shorter wavelength systems, the effect is typically rather small. But, for longer wavelength systems, the ionospheric delay can be significant, especially during strong solar activities with up to 100 TECU. Again, the effect on InSAR is only relevant for differences in the path delay between pixels and the electron content in the ionosphere only changes gradually in space, so that the effect is rather small on interferograms, but can become significant for long wavelengths.

After the ionosphere, the signal is passing the troposphere. In the troposphere, the main factor for the signal path delay is the water humidity content in the atmosphere. This is not constant but changing in time and space, and affects all wavelenghts. If we imagine clouds, we understand that the humidity is different inside the cloud and outside and that clouds are spatially and temporally changing. Changes in humidity and pressure are significant in terms of interferometry and can have a strong influence on the phase differences. The influence normally gets stronger for pixels that are far away from each other, whereas an influence on the phase difference from the atmosphere can typically be neglected for neighboring pixels. This is the case because the atmosphere and the water content of the atmosphere gradually changes and we typically do not see sudden jumps in the atmospheric phase delay. Extreme weather conditions in a turbulent atmosphere may, however, cause more extreme differences in the phase delay.

The atmospheric delay, or rather the changes in the atmospheric delay between acquisitions, also known as atmospheric phase screen

(APS), is a large error source in SAR interferometry. It puts the standard textbook sentence on the weather independence of SAR into question. Although we can get SAR images under (almost) all weather conditions, measurements in and with SAR images are always influenced by the atmosphere.

Methods for removing the influence of the atmospheric delay are available. One way is to reduce the time difference between the interferometric acquisitions close to zero, as it is the case in bi-static SAR missions. Another way is to use additional information from global weather models to estimate and remove the path delay, which will be discussed in Chapter 10. Finally, having several images, the path delay can be estimated from the data itself and removed, which will be discussed in Chapter 6.

3.4.4   Orbit Errors

The position of the sensor needs to be estimated at high precision to allow for correct baseline estimation. Furthermore, changes in the baseline between images forming an interferogram can become problematic and need to be corrected. All of this requires a precise knowledge of the sensor position at all times. For spaceborne systems, this is the orbit of the satellite and therefore the term "orbit errors". The problem is generally even larger for airborne systems due to the greater variation in a flight path during acquisition. Errors in the orbit estimation of spaceborne sensors typically lead to rather distinct patterns in the resulting interferogram, the so-called phase-ramps. They are recognizable by linear changes in the phase. It is very uncommon to find linear patterns in natural topographies, deformations, or atmospheric influences, making errors from orbit estimations relatively easy to distinguish in the interferograms. Often, two sets of orbit information are available for spaceborne SAR sensors. A standard orbit, sometimes also called rapid orbit, is the orbit estimation directly available during or after image acquisition. This orbit information is normally rather imprecise and not well suited for measurement within

SAR images. A more precise orbit is calculated later, based on the measured path of the satellite. This is called precise orbit and becomes available a few days after image acquisition. For non-time critical InSAR applications, the general advice would be to always use the precise orbit, drastically reducing orbit errors.

3.5 Conclusions

SAR interferometry is a very precise method for DSM generation. With the high precision, various error sources have to be considered. Many of these errors can be reduced or avoided when the time difference between the acquisitions is reduced, ideally to zero, avoiding atmospheric differences and temporal decorrelation. The wrapping of the phase and the inherent ambiguity of the phase measurement cannot however, be avoided in this way. Multiple acquisitions with different baselines can solve the ambiguities in this case.

Nevertheless, SAR interferometry is a complicated multi-step process with several possible error sources. It therefore should be processed carefully with quality assurance measures to assure that the derived DSM is correct. If processed correctly though, InSAR allows for large area DSM generation in high precision. This is the reason why the most well-known and precise global DEMs, i.e. the SRTM and the TanDEM or WorldDEM, are generated from InSAR data.

Chapter 4

StereoSAR

SAR interferometry offers unique capabilities in DEM generation with respect to precision and wide area coverage. However, although SAR interferometry has great potential, it suffers from temporal decorrelation and atmospheric disturbances, as explained in the previous chapters. These are actually not small problems, but are so severe that repeat-pass SAR interferometry cannot be used in large areas of the world, as thick vegetation coverage quickly leads to decorrelation. The best solution to this problem is to reduce the time difference between the acquisitions, also called the temporal baseline, as much as possible. With bi-static configurations, the temporal baseline is reduced almost to zero. However, this requires a bi-static system, which is significantly more expensive and data availability might be restricted. Another possibility for digital surface model (DSM) generation with spaceborne data is using photogrammetry with optical data. Good precision can be achieved with optical data as well. However, in several areas of the world, especially around the tropics, almost permanent cloud coverage hinders the usage of optical data, whereas the dense vegetation leads to fast temporal decorrelation, rendering SAR interferometry useless. Under these circumstances, SAR offers another possibility for height estimation using the dependency of the pixel positioning in range direction on the height of an object. Similar to photogrammetry, in SAR radargrammetry, the geometrical properties of the SAR imaging properties are used to measure topographical features.

Stereo-radargrammetry was the first method used to derive DSMs from airborne radar data. La Prade (1963) defined the first principles of stereo radargrammetry, which later became operational (Rosenfield, 1968). Leberl *et al.* (1986) showed radargrammetry with spaceborne data. As the precision of radargrammetric DSMs depends on the spatial resolution, radargrammetry had a renaissance after the availability of commercial high-resolution SAR data. Radargrammetry was demonstrated with high-resolution TerraSAR-X data (Raggam *et al.*, 2010), COSMO-SkyMed spotlight data (Capaldo *et al.*, 2011), as well as with Radarsat-2 data (Toutin & Chenier, 2009).

4.1   Introduction to Stereo-Radargrammetry

In stereo radargrammetry, two SAR images acquired from different orbits are used to determine the relative height of points by measuring the difference in the range shift, also called parallax, which depends on the incidence angle difference between the images $\Delta\theta$.

$$\Delta\theta = \theta_1 - \theta_2 \tag{4.1}$$

The basic principle is shown in Figure 4.1. Depending on the incidence angle, objects will be mapped in range direction according

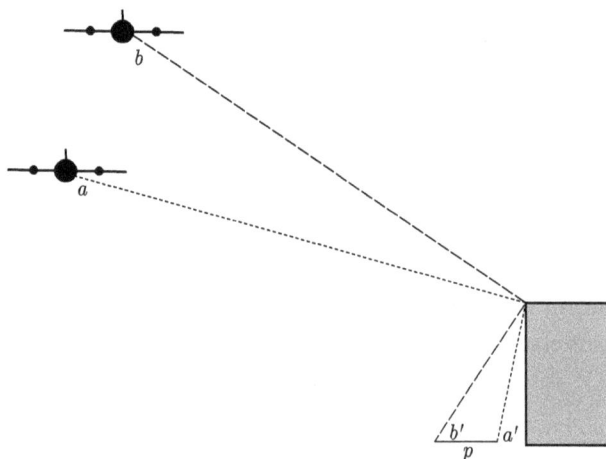

Figure 4.1   SAR configuration for stereo radargrammetry.

to their height. This shift is relatively small for images with a flat angle, but increases with images acquired from steeper angles. This allows for a reconstruction of the height when homologous points can be found in both images and the range-shift difference can be determined. As the extension of the range shift depends on the relative height of the object, this can be used to derive relative heights. The technical challenge in this process is the identification of homologous points in both images.

4.2   Finding Homologous Points

Stereo radargrammetry depends on the precise identification of homologous points in both images. There are several approaches available. The standard approach would be using the normalized cross-correlation, searching for the highest cross-correlation σ from (see Fayard *et al.*, 2007)

$$\sigma = \frac{1}{N-1} \sum_{x,y} \frac{(m_{x,y} - \bar{m})(s_{x,y} - \bar{s})}{\sigma_m \sigma_s} \tag{4.2}$$

with N representing the number of valid pixels in the search window, and $m_{x,y}$ and $s_{x,y}$ representing the pixel value at (x, y) in the search window of the master and the slave, respectively. $\bar{m}$ and $\bar{s}$ are the mean values, and σ_m and σ_s are the standard deviation of the values in the search window of the master and slave images. The normalized cross-correlation is stable and efficient. We saw this method already in Section 3.2.3.1 of Chapter 3.

Other methods, like, for example, Mutual Information (Viola & Wells, 1997; Xie *et al.*, 2001) or Semi-Global Matching (Hirschmuller, 2007) could also be used. Edge-based matching is also a common approach and it has also been shown that a combination of edge- and gray-level image matching can achieve good results as well (Paillou & Gelautz, 1999).

A common problem in the process is false matching results. One way to reduce false matching is by applying constraints on the image matching, for example, by using an epipolar geometry. An epipolar image is defined as having the slant-range shift only occurring along the range axis of the image, so it requires co-registration in azimuth. This is possible because the position in azimuth does not depend on the height of the object. However, strong topographic effects may require correction as well, for example, based on a low-resolution DEM, as do orbit effects. After creating an epipolar image, the homologous points in both images need only to be searched in range direction, making the process faster, and if the epipolar image has been created correctly, it is more accurate and reliable.

Hierarchical matching strategies, such as the pyramid procedure, can also be used (Denos, 1992). The idea is simple. An image pyramid is built, starting with the original image. At each level going up, the image size is reduced by a factor of 2^n at the nth iteration. The image matching starts at the top of the pyramid, with the lowest resolution images and works down the pyramid, increasing the resolution at each step.

Working with SAR images, the negative effect of the speckling has to be taken into consideration. A speckle filter is normally applied before searching the homologous points to provide more stable results that are less influenced by the speckle noise. You can refer to Section 2.5.2 of Chapter 2 for more information on available speckle filtering approaches.

4.3   Deriving Relative Height Information

The height can be calculated with

4.1. Height from StereoSAR:

$$h = \frac{p}{\cos\theta_1 \pm \cos\theta_2} \tag{4.3}$$

---

where h is the relative height, p is the range difference in slant range, and θ_1 and θ_2 are the incidence angles. This can be the sum of the cosines of the angles or their difference, depending on the stereo configuration; they are added in the standard same-side configuration, as shown in Figure 4.1, and subtracted if an opposite-side configuration is used. Although it could theoretically achieve higher accuracy, the opposite-side configuration is seldom used, as it is very difficult to find homologous points in such a configuration. As we can see from this equation, the achievable precision depends on the difference between the image incidence angles, where larger incidence angle differences allow for a more precise measurement of the height. However, larger difference leads to more differences in the images, making it again harder to find the homologous points.

4.4   DSM Generation

After the relative heights of each point are calculated, it is necessary to convert the relative heights to absolute heights, for example, by establishing a reference point with a known height or by minimizing the absolute difference between a known DEM and the newly created one, and hopefully more precise, radargrammetric DSM. Based on these absolute height values, each pixel can then be transformed into world coordinates to create a final DSM. Since the points will not form a uniform grid after conversion, it is necessary to further interpolate them to form a uniform grid, for example, by triangulating between the georeferenced points and estimating the grid height based on the triangulation. Other interpolation methods, like Kriging, etc, can also be used. Often, the final DSM contains void areas, either from radar shadows and/or from areas of very low cross-correlation values that may, and normally should, have been masked.

4.5   Deriving Absolute Height Information

Alternatively, the absolute position can be directly derived from stereo-radargrammetric observations. The geocentric coordinate

(x, y, z) can be derived from an oversized system of equations with (X, Y, Z) being the satellite position at image 1 and image 2, and $(\mathbf{X}_v, \mathbf{Y}_v, \mathbf{Z}_v)$ being the velocity vector of the satellite at images 1 and 2:

$$
\begin{cases}
r_1^2 = (x - X_1)^2 + (y - Y_1)^2 + (z - Z_1)^2 \\
0 = (x - X_1)^2 \mathbf{X}_{v1} + (y - Y_1)^2 \mathbf{Y}_{v1} + (z - Z_1)^2 \mathbf{Z}_{v1} \\
r_2^2 = (x - X_2)^2 + (y - Y_2)^2 + (z - Z_2)^2 \\
0 = (x - X_2)^2 \mathbf{X}_{v2} + (y - Y_2)^2 \mathbf{Y}_{v2} + (z - Z_2)^2 \mathbf{Z}_{v2}
\end{cases}
\tag{4.4}
$$

With three unknowns and four equations, a solution can be found, but nevertheless, errors have to be taken into consideration. Errors stem from inaccurate matching or inaccurate sub-pixel matching, orbit errors of the satellite metadata and incorrect estimations of the range r, especially those from atmospheric path delay differences. This will be discussed in more detail in Chapter 10.

4.6   Example Application of StereoSAR at Mount Song

The example uses TerraSAR-X stripmap data covering Mount Song in China's Henan province. Mount Song is also famous outside of China for having the Shaolin Monastery located on the mountain. Mount Song itself is not very high, only reaching about 1500 m, but the slopes are steep. The mountainsides are densely vegetated, causing quick temporal decorrelation when using interferometric SAR. StereoSAR is less affected by this though.

We use a pair of TerraSAR-X stripmap images acquired from ascending orbits. The amplitude data of the two images are shown in Figure 4.2.

A very heavy rainfall event occurred during the acquisition of the data on 2011-07-12, as can be seen in Figure 4.2(a). Despite the claim that SAR is independent of weather, very heavy rainfall can disturb SAR images, especially when using shorter wavelengths. These events clearly will negatively influence the StereoSAR results. Nevertheless, a

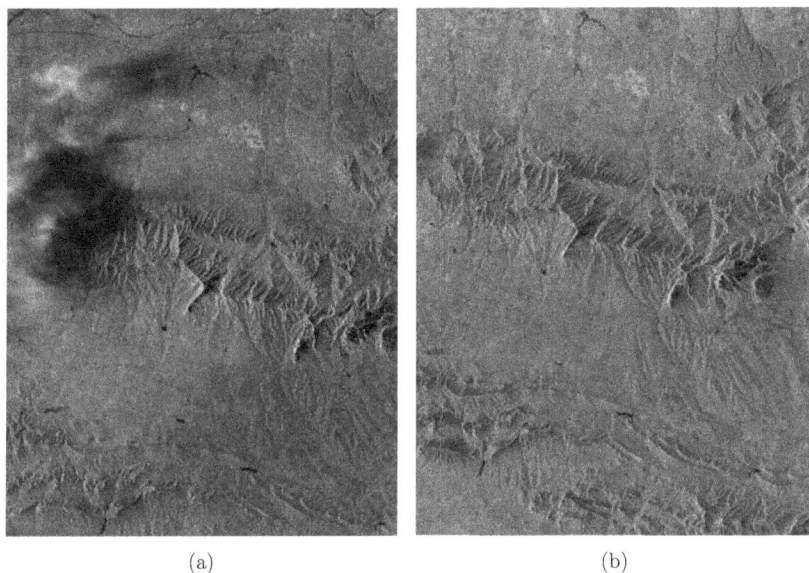

(a)   (b)

Figure 4.2 TerraSAR-X strimap image of 2011-07-12 (a) and 2011-07-18 (b) used for the StereoSAR experiment (© DLR, 2011).

stable DSM can still be generated from areas not covered by heavy rainfall. Such extreme rainfall and the accompanying turbulences and water vapor differences in the atmosphere will, however, significantly influence SAR interferometry, so that no usable DSM will be retrievable under such circumstances.

StereoSAR is therefore a much more stable technique. However, the precision of the results depends on the spatial resolution of the data, with higher spatial resolution providing more precise results. Furthermore, interferometric SAR can deliver much more precise results than StereoSAR under optimal circumstances. However, such 'optimal circumstances' can typically only be realized with bi-static configurations. Nevertheless, a much wider range of datasets can be processed given the stability of the StereoSAR approach.

The results are shown in Figure 4.3. The DSM can be seen in Figure 4.3(a) and the maximum correlation is shown in Figure 4.3(b). We can see the negative effect of the rainfall on the outcome from the maximum correlation because no matching points with a high correlation are found

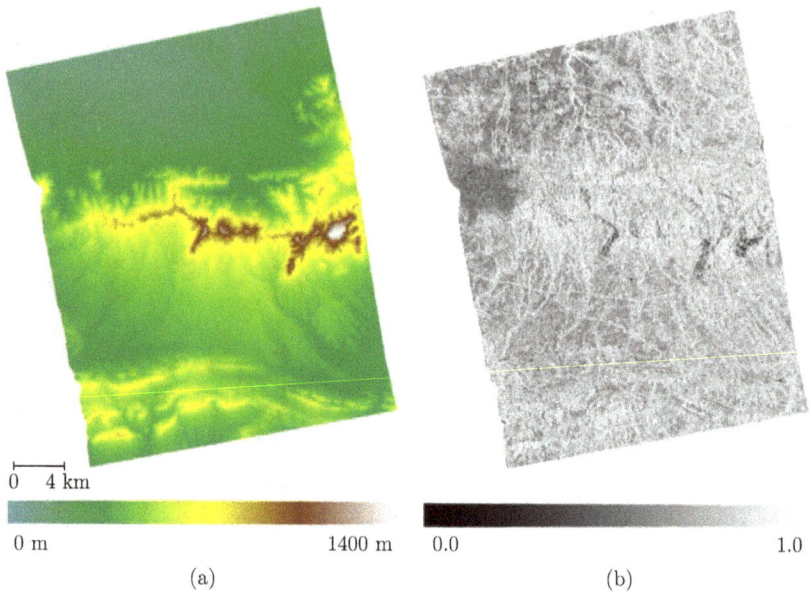

0 m   1400 m   0.0   1.0

(a)   (b)

Figure 4.3   Derived DSM using StereoSAR (a) and maximum correlation (b).

in this area. Nevertheless, in many areas, the points can be matched with a comparably high correlation.

The overall absolute mean error of the accuracy is about 6.2 m with a root mean square of 10.4 m of the error (Balz *et al.*, 2013). In a similar experiment with TerraSAR-X spotlight data, the accuracy was improved to a 8.7 m root mean square error, clearly demonstrating the strong dependence of the achievable accuracy on the spatial resolution. The results in Balz *et al.* (2013) are based on a deviation of the StereoSAR method, but the results are quite similar to what is generally expected from the standard StereoSAR approach.

4.7   Conclusions

Stereo-radargrammetry is a very stable and (almost) weather-independent process. As it relies on the amplitude instead of the phase, it is much more resilient against atmospheric distortions and temporal decorrelation. It is a standardized process, comparable to photogrammetry, with the possibility of using some of the well-established photogrammetry workflows, making it very suitable for use in environments with good photogrammetric experience. Stereo-radargrammetry relies on the accurate identification of homologous points though and this requires a certain identifiability of points. It therefore works well in images with a lot of structural elements that permit easy identification of homologous points, but may fail in areas with extremely homogenous surfaces. Unfortunately, this may even be the case in our application, as some tropical forests present a very homogenous surface on the top of the canopy. Here, radargrammetry may also not achieve the desired results. A further drawback is the dependence of the achievable accuracy from the spatial resolution of the system. The points are found in pixel precision or sub-pixel precision. So, the accuracy directly depends on the resolution. This is a big difference to the interferometric approach, where the theoretically achievable accuracy depends on the wavelength of the system, not the spatial resolution. As the effective spatial resolution is further reduced by the necessary speckle filtering, a good compromise for the size of the speckle filter has to be found, so that the speckle is effectively reduced, while the spatial resolution is reduced only minimally.

Part III   Surface Motion

Chapter 5

Differential SAR Interferometry

In differential SAR interferometry (D-InSAR), the basic InSAR principles apply. As explained in Chapter 3, we use the phase difference between two SAR acquisitions:

$$\phi = \phi_1 - \phi_2 = \frac{4\pi}{\lambda}\Delta r \qquad (5.1)$$

where ϕ_1 and ϕ_2 are the phases of the two acquisitions, λ is the radar wavelength and Δr is the difference in range between the two acquisitions (Figure 5.1). For DEM generation, we use the difference in range between the acquisitions to derive the relative height of a point via their trigonometric relation. However, differences in range between acquisitions can also be caused by a motion of the point between acquisitions. Previously, we discussed the effect of moving targets under the viewpoint of noise and loss of coherence. However, slow-moving targets can keep the phase coherent between acquisitions, allowing for a measurement of the difference in range caused by the motion. This requires the motion between the acquisitions to be less than half of the system's wavelength. With a typical repeat cycle of spaceborne SAR systems between 10 and 30 days and a wavelength between 3 and 20 cm, only slow surface motion can be measured with this method. Surface motions in the range of a few centimeters per year can be found, for example, in tectonic motions, urban subsidence, slow-moving landslides, volcanic activities, etc. These phenomena are difficult and costly to continuously survey by other means, while radar remote sensing offers wide area coverage and precise measurement of

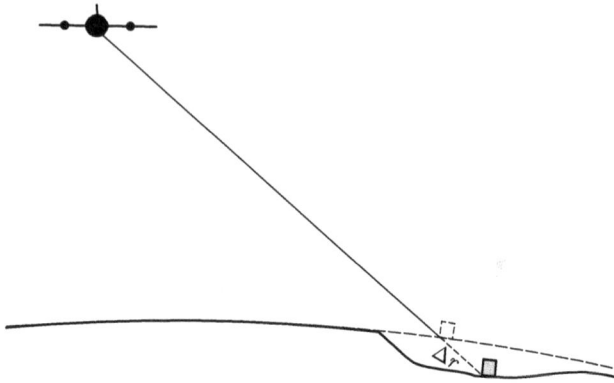

Figure 5.1   D-InSAR geometry with Δr being the motion (or range difference) in the line-of-sight direction.

these slow surface motions. However, there are several noise factors to be considered. First, the temporal decorrelation: This is the main disturbing factor. In practice, we go to great lengths to avoid it, even launching bi-static satellite missions to allow for a generation of DEMs without interference from temporal decorrelation. However, this does not work when measuring motion. To measure motion, there needs to be a significant time difference between the acquisitions, otherwise there is no motion signal to be measured. With such a time difference, unavoidable when interested in measuring motion, the problem of temporal decorrelation is unavoidable. Furthermore, with a time difference between the acquisitions, the atmosphere and weather situation also changes, including differences in the atmospheric path delay, which again disturbs the motion measurements.

The interferometric phase ϕ of a stable object therefore consists of several contributions:

5.1. Phase contributions:

$$\phi = \phi_{\text{flat}} + \phi_{\text{topo}} + \phi_{\text{motion}} + \phi_{\text{atmo}} + \phi_{\text{orbit}} + \phi_{\text{noise}} \qquad (5.2)$$

---

where ϕ_{flat} is the phase contribution caused by the range differences at 0 height, the so-called flat Earth, ϕ_{topo} is the phase contribution from the topographic height and the baseline difference between the acquisitions. This is the phase contribution, as we recall, used to generate a DEM. ϕ_{motion} is the phase contribution caused by the motion and the element of the interferometric phase we are concerned with in differential SAR interferometry. ϕ_{atmo} is the phase contribution from the atmosphere caused by differences in the atmospheric phase delay between the acquisitions, also called atmospheric phase screen or APS. ϕ_{orbit} is the phase contribution caused by inaccuracies in the orbit determination of the sensors. Other noise sources are aggregated in ϕ_{noise}, which, for example, includes the thermal noise of the system itself. ϕ is the phase of a stable object. For a clear determination of ϕ_{motion}, it is important to either know the other phase terms or to reduce them toward 0 as much as possible. There are two main ways of removing ϕ_{topo} in D-InSAR: using a DEM for two-pass D-InSAR and using an additional image for three-pass D-InSAR.

5.1   Two-Pass D-InSAR

In two-pass D-InSAR, two images are used to determine the motion element between the acquisition. There needs to be a significant time difference between the images to allow for a motion to be significant enough to be measured. The required time for measurement depends on the velocity. In typical repeat-pass D-InSAR, each pass of the satellite is separated by several days, reducing the measurable velocities to very slow-moving objects, such as surface motion from subsidence, tectonics, and volcanic activities. The contribution from the topographic height ϕ_{topo} is removed based on a pre-existing DEM. This process is identical to the topographic phase removal described in Chapter 3. As described in Chapter 3, this requires the DEM to be transformed into the SAR coordinate system first, which is also described in detail in Chapter 3.

The phase for each point is calculated based on

$$\phi_{topo} = \frac{4\pi h}{\lambda} \frac{B_\perp}{r \cdot \sin\theta} \tag{5.3}$$

where h is the height above the estimated zero height from the flat-Earth removal. The calculated phase ϕ_{topo} is then removed and we then assume that the remaining $\phi - \phi_{topo} = \phi_{motion}$, ignoring the other phase contributions.

D-InSAR does ignore large phase contributions, especially from the differences in the atmospheric phase delay between acquisitions. This can significantly influence the D-InSAR results. Furthermore, ϕ_{topo} estimated on the DEM is itself often not very precise due to errors in the DEM. There remains an influence from the topography, the topographic error, which is the phase residual remaining from an incorrect removal of ϕ_{topo}.

5.2   Three-Pass D-InSAR

Three-pass D-InSAR attempts to improve the estimation of ϕ_{topo} by including an additional SAR image, using one pair to estimate the motion ϕ_{motion} and another pair to estimate ϕ_{topo}. This can reduce errors arising from an imprecise DEM in the two-pass D-InSAR approach. However, this requires an additional image. Even more so, it requires an image pair without or with only a minimal ϕ_{motion} to derive ϕ_{topo} from, otherwise the motion component also influences the estimation of ϕ_{topo}. So, three-pass D-InSAR works well if we find a singular motion event, for example, an earthquake, allowing the use of a pre- or post-earthquake data pair to estimate ϕ_{topo} without the motion component from the earthquake and then estimate ϕ_{motion} on the co-seismic image pair, which is an image pair with one image acquired before and one image acquired after the event, thus measuring the motion caused by the event. Furthermore, the pair estimating ϕ_{topo} should have a large perpendicular baseline, whereas the pair estimating ϕ_{motion} should have a minimal perpendicular baseline to minimize the effects of the topographic phase. For continuous velocities, three-pass D-InSAR has the problem that the motion is influencing all image pairs, which hinders the clear separation of ϕ_{topo}.

Another way to reduce this is by having image pairs with different temporal and spatial baselines, where one pair with a very short temporal and a large spatial baseline is used to estimate ϕ_{topo}, whereas a second pair with longer temporal and short spatial baseline is used to estimate ϕ_{motion}, as the overall motion will be much larger for images with longer time difference between them.

For the three-pass D-InSAR to work correctly, the topographic phase should be very prominent in the topographic interferogram and the deformation phase should be dominant in the interferometric pair used to represent the deformation. For this, the difference in the perpendicular baseline between the interferograms is also very important. Furthermore, the topographic phase in the three-pass D-InSAR should have been unwrapped. The unwrapped topographic phase ϕ_{topo} is normalized by the ratio between the perpendicular baseline of the topographic interferogram $B_{\perp,topo}$ and the perpendicular baseline of the deformation interferogram $B_{\perp,defo}$. This phase is subtracted from the phase of the flattened deformation interferogram φ_{defo}.

$$\varphi = \varphi_{defo} - \frac{B_{\perp,topo}}{B_{\perp,defo}} \cdot \phi_{topo} \qquad (5.4)$$

5.3 Interpretation of Wrapped Differential Interferograms

The interferometric phase caused by the motion can be calculated as

5.2. Motion-related phase:

$$\phi_{motion} = \frac{4\pi}{\lambda} \Delta r \qquad (5.5)$$

where Δr is the motion in the line-of-sight (LOS) direction of the system. ϕ_{motion} is a wrapped phase though. That is to say, for motions larger than half of the system's wavelength, the wrapped ϕ_{motion} is ambiguous. It is therefore necessary to unwrap the phases for a precise estimation of the motion component. This is described in the following section. Nevertheless, even without unwrapping, differential interferograms can be interpreted in search for deformations, which is often sufficient if one is not interested in the precise deformation value.

The fringes in the image show the deformation. Similar to fringes in height estimation, where denser fringe patterns show steeper topography, in D-InSAR, denser fringes relate to faster motion. When interpreting the results, one needs to keep in mind that height estimation errors also lead to fringes in the resulting differential interferogram as well as atmospheric effects. This makes the interpretation of differential interferograms difficult.

5.4   Measurement with Unwrapped Differential Interferograms

To measure deformation, the interferometric phases need to be unwrapped. Again, like the topographic phases, the differential phase information is also relative and shows us the difference in motion between two pixels. Unwrapping is similar to the phase unwrapping for topographic height estimation. From differences in the successfully unwrapped phase, motion differences between pixels can be calculated using

$$\phi = \frac{4\pi}{\lambda} v_{los} \tag{5.6}$$

$$\phi_1 - \phi_2 = \frac{4\pi}{\lambda}(v_{los,1} - v_{los,2}) = \frac{4\pi}{\lambda}\Delta v_{los} \tag{5.7}$$

5.5 Example: Coseismic Interferometry for the Bam Earthquake

On December 26, 2003, according to the US Geological Survey, an earthquake of 6.7 magnitude shook the small town of Bam in Iran.

We selected the image from 2003-12-03, acquired before the earthquake, as master image. The image is shown in Figure 5.2. As the original ASAR images do not have square pixels, the image in Figure 5.2 has been multi-looked four times in azimuth to correct for this limitation. The interferograms presented in the following have also been multi-looked accordingly.

Figure 5.2 ASAR amplitude image from Bam, Iran, acquired on 2003-12-03.

The first step in the interferometric processing is the co-registration of the images. In this example, a coarse-to-fine strategy, as described in Chapter 3, is used. After resampling, the quality has to be checked.

The interferogram is formed after resampling by the complex conjugate multiplication of the two SAR images, which forms the difference between the phases of the two images. Afterwards, we remove the phase component from the flat Earth. The resulting interferograms are shown in Figure 5.3(a). The flat-Earth removed interferogram in Figure 5.3(a) is still dominated by the topographic phase components, so that the earthquake-induced motion cannot be clearly identified.

5.5.1 Two-Pass D-InSAR Example

As discussed previously in this section, one way to form a differential interferogram is the two-pass method, where the topographic phases are removed by simulating the expected phases from a DEM. In our example, we used the SRTM v4 DEM of Bam, Iran. The simulated phases were removed from the interferogram, leading to a two-pass differential interferogram as shown in Figure 5.3(b).

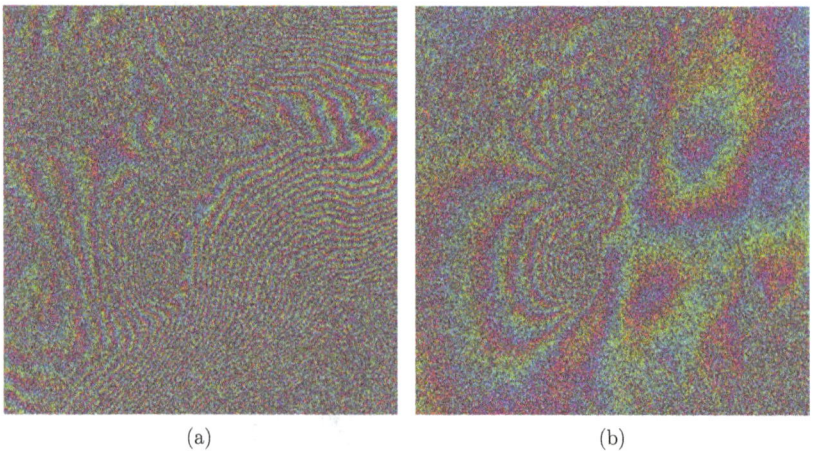

(a)   (b)

Figure 5.3   ASAR interferogram of 2003-12-03 and 2004-01-07 after flat-Earth removal (a) and two-pass differential interferogram (b).

In Figure 5.3(b), the motion of the earthquake is now visible. Ignoring noisy image parts, the motion can be visually interpreted and quantitative statements can be formulated by counting the wrapped phase cycles along the motion line. This already can in many cases be enough for a motion analysis.

5.5.2 Three-Pass D-InSAR Example

Three-pass D-InSAR requires two interferograms, one containing the topographic phase component and ideally no motion, while the second interferogram should be dominated by the motion-induced phase. The correct selection of the temporal and perpendicular baselines is therefore very important. For the three-pass D-InSAR, we form the topographic phase from the images on 2003-12-03 and 2003-06-11, with both images before the earthquake and with a relatively large perpendicular baseline of −475 m. We then use the interferograms of 2003-12-03 and 2004-01-07 as well as 2003-12-03 and 2004-02-11 to get the phase from the earthquake-induced motion. As shown in Table 5.1, the interferogram with the image on 2004-01-07 has a short temporal baseline, which is good to avoid temporal decorrelation and also a rather long perpendicular baseline of 520 m. The temporal baseline for the interferogram between 2003-12-03 and 2004-02-11 is longer, but the perpendicular baseline is very short.

The results of both three-pass interferograms are shown in Figure 5.4. This image demonstrates the importance of the perpendicular baseline selection, as we get a very clear picture of the earthquake-induced

Table 5.1   SAR data used for the example.

Sensor	Date	Master	Baseline (m)
ENVISAT ASAR IMS	2003-06-11	S	−475
ENVISAT ASAR IMS	2003-12-03	M	0
ENVISAT ASAR IMS	2004-01-07	S	520
ENVISAT ASAR IMS	2004-02-11	S	−1.7

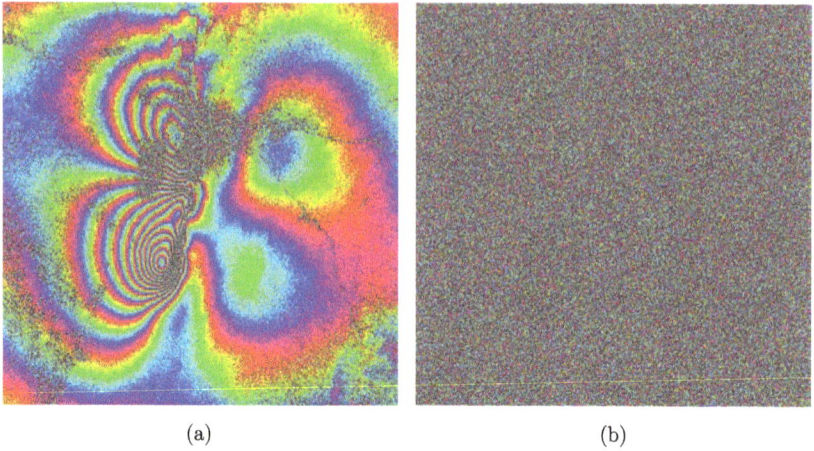

(a)   (b)

Figure 5.4   ASAR three-pass D-InSAR interferograms with the deformation phase between 2010-12-03 and 2004-02-11 (a) and 2010-12-03 and 2004-01-07 (b).

motion in Figure 5.4(a), but mostly noise in Figure 5.4(b). With the perpendicular baseline even longer in the deformation interferogram, the three-pass D-InSAR approach cannot remove the topographic phase component correctly and the results are meaningless in Figure 5.4(b).

5.6   Conclusions

Differential SAR interferometry is a powerful technique for measuring surface motion. However, it suffers from numerous noise sources. The differential phase contains not only the deformation information but also the topographic height error, phase from atmospheric path delay differences, orbit errors, and noise. Furthermore, temporal decorrelation makes differential SAR interferometry unusable in areas with dense vegetation cover. Overall, the technique is often not usable, and successful application remains possible mostly in areas with limited vegetation coverage. The following chapters describe methods for improving D-InSAR, so it can also be widely used for long-term monitoring of surface motion.

Chapter 6

Permanent Scatterer Interferometry

Permanent Scatterer Interferometry (PSI) describes various approaches for overcoming the limitations of D-InSAR using a stack of time-series SAR images. A stack of time-series SAR images is a series of SAR images, covering an identical area suitable for SAR interferometry. Therefore, the images need to be acquired from a similar orbit, with a similar sensor and similar acquisition parameters to allow for interferometric processing. One advantage of SAR processing is that it is relatively easy to collect such stacks, due to the almost independence of SAR to weather conditions and the geometric simplicity of SAR imaging. Unlike optical imaging, there are no lens distortions or atmospheric distortions that hugely affect the geometry of an SAR image. The goal of PSI is the estimation of slow surface motions over long periods.

As an extension of D-InSAR, the various PSI methods aim to overcome temporal decorrelation and atmospheric effects in InSAR processing. In terms of atmospheric effects, we have to separate the influence on the SAR amplitude image and the phase. For the amplitude image and the image geometry, only under extreme weather conditions we can see an effect, stemming from the path delay. For interferometry, we measure fractions of a wavelength in millimeters or centimeters; therefore, even small delay differences have a direct effect on an interferogram. Such effects appear in almost all interferograms. So, when we say that SAR images can be acquired under (almost) all weather conditions, this does not negate that the atmosphere still influences the electromagnetic signals throughout an image and between different images. The first PSI approach was PSInSAR, appearing around the year 2000 (Ferretti *et al.*, 2000, 2001). Afterwards, other approaches followed, for example, STUN (Kampes, 2006), SBAS (Crosetto *et al.*, 2005), or StaMPS (Hooper *et al.*, 2004), which is currently probably the most widely used approach. As all these approaches are based on measuring point-like stable targets, the so-called Permanent Scatterers (PS), we group them all together under the general term PSI. Other techniques, like SBAS (Berardino *et al.*, 2002), QPS (Perissin & Wang, 2011), or SqueeSAR (Ferretti *et al.*, 2011) do not rely exclusively on these PS points. Sometimes, these approaches are also categorized under PSI, sometimes they are described as InSAR time-series analysis or using other terms. PSI is a complex, multi-step approach. In the following section, we follow the steps with a particular focus on the standard PSInSAR approach.

6.1   Identifying the Permanent Scatterers

In the center of Permanent Scatterer Interferometry (PSI) are the Permanent Scatterers (PS). Permanent Scatterers, also known as Persistent Scatterers, are point-like targets in an SAR image that stay stable over long periods of time. They are defined as having one dominant scatterer in a resolution cell. They are typically formed by dihedral or trihedral structures, like building edges, balconies, etc., or poles, like street lamps (see Perissin & Ferretti, 2007). Dominant and point-like scatterers have some big advantages making it easier and more reliable to analyze them. If the backscattering from one resolution cell is dominated by one single scatterer, there is no speckling effect for that resolution cell. Speckling is the result of the interference of the backscattering of multiple scatterers, which does not happen if the signal in one resolution cell is dominated by the response of one single scatterer. Without speckling, the signal from this resolution cell can be considered a deterministic signal in contrast to the probabilistic signals received from resolution cells suffering from speckling. Another advantage of PS is the absence of temporal decorrelation. PS stay stable over long periods. Again, this is because they are typically formed from dihedral or trihedral structures or poles. Such structures do not move (much) with the wind or grow like, e.g. vegetation does. They remain stable, offering a stable deterministic signal over long periods. Analyzing only PS gives us the opportunity to work with deterministic signals that do not suffer from temporal decorrelation and thus ideal for long-term deformation monitoring.

The first task in this approach is finding such PS in the data. One method, already described in the first PSInSAR approach is the amplitude dispersion index (Ferretti *et al.*, 2000, 2001). For PS, we are interested in their phase stability. As shown by Ferretti *et al.* (2001), the phase dispersion can be estimated from the amplitude dispersion of a pixel at least for pixels with high signal-to-noise ratio (SNR). This is a practical and fast way to estimate phase stability, as it only requires the amplitude of the SAR images. The amplitude dispersion index D_A of a pixel is calculated with

$$D_A = \frac{\sigma_A}{\overline{m}_A} \tag{6.1}$$

where $\overline{m}_A$ is the mean value of the amplitude of a given pixel in all co-registered SAR images forming the SAR stack used for PSI and σ_A is the standard deviation of the stack. Based on this, the so-called PS candidates (PSC) are selected, as a low amplitude dispersion index is an estimation for a low phase dispersion and therefore an indication that a pixel contains a PS. At this stage, this is just an estimation and therefore these points are called PSC. PSC are selected based on a threshold, typically with $D_A < 0.25$, but this can vary based on the available SAR data and the area of interest.

Selection of PSC based on the amplitude dispersion index is the standard method because it offers fast results that are suitable for high SNR in most cases. Thus, for strong reflectors, as we can find them in large numbers in urban areas, the amplitude dispersion index is fast and sufficiently accurate to estimate a large number of suitable PSC for further processing. Nevertheless, there are other methods available.

For example, a high signal-to-clutter ratio can be used (Kampes, 2006), which leads to rather similar results.

One critique on using the amplitude dispersion index is its validity for high SNR pixels only. However, in areas outside of cities, we may find stable points that have a lower SNR, as these PS may be formed from, e.g. exposed rocks that can provide phase stability over long periods, but do not have a high SNR and may therefore not be considered. In the StaMPS approach (Hooper *et al.*, 2004), PSC are initially selected with a less restricted threshold, e.g. $D_A < 0.4$, or higher. A series of processing steps follow for estimating phase stability of these PSC. Actually, StaMPS spends most of its processing time on the identification of the PS in the data. More on StaMPS can be found in Section 7.3 of Chapter 7.

Selecting the PSC is just the first, but a very crucial step. The following processing steps are based on these selected PSC. It is therefore important that enough PSC are found all over the image to ensure adequate coverage and a high enough PS density to estimate atmospheric effects.

Working with PSCs from now on also means that we leave the realm of 2D image processing and enter the world of point clouds.

6.2   Pre-Processing Steps

PSI is a complex set of processing steps. Before it even starts, a series of interferometric steps are necessary. First, a stack of SAR images is collected and co-registered to a single master image. In the standard PSI approach, all interferograms are also formed with respect to this master; however, this is not a given for all approaches. After co-registration, the selection of the PSC via the amplitude dispersion index is often the next step. However, it is also possible to process the interferograms for the complete images and select the PSCs later on.

Ideally, the master image is selected in a way that minimizes the perpendicular and temporal baselines. Furthermore, weather information can be added to avoid strong rainfall in the master image, as this would include turbulences and therefore probably large atmospheric phase screen (APS) differences throughout the image. Such turbulences in the master image would affect all interferograms, as in a single-master configuration, all other images will interact with the one master image.

After the co-registration, the PSCs can already be selected based on the amplitude dispersion index. This saves processing time in the processing steps, as the following steps can only be done for the PSC, which consists only of a fraction of the overall image pixels.

For all PSC, the interferogrammetric information after removal of the topographic phase is calculated based on a DEM. With the almost global availability of the SRTM DEM and the global availability of the TanDEM DEM, several options are available. Additionally, it is also possible to create a DEM via interferometric processing for use in further processing.

During the later processing steps, a residual topographic error, or DEM error, is estimated for each PSC. The DEM does not need to be perfect. A minimization of the residual topographic errors by choosing the best available DEM is still the preferable option.

6.3   Generation of A PS Network

From the available PSCs, a subset is selected to form a network of points. This network is used later on to estimate the atmospheric phase screen (APS). To form a network that covers the complete image, the most stable points are selected, which are often the points with the lowest amplitude dispersion index. Afterwards, points in close proximity to each other are removed, so that the remaining points cover the image at a minimum distance between each other. This can also be achieved by sub-dividing the image into grid cells and selecting the best point in each grid cell. At the end, a sparse network of the most stable points should completely cover the image (Figure 6.1). In many cases though, it is often not possible to cover the complete image, as there might be areas containing no stable points or not enough stable points. This can lead

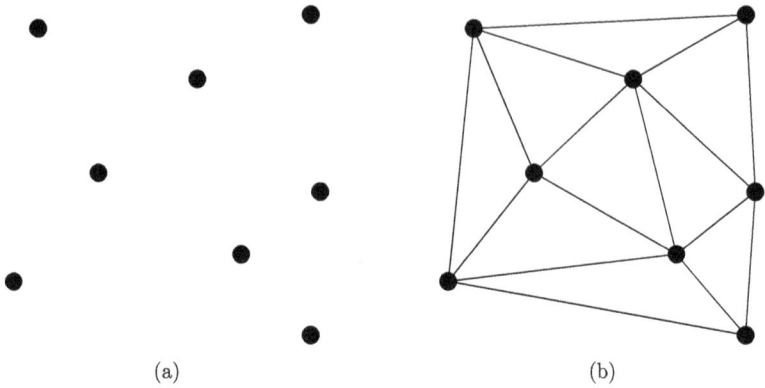

Figure 6.1 Selected sparse PSC points (a) and connected PS points using Delaunay triangulation (b).

to difficulties in estimating the APS in those areas later on, leading to errors in the velocity estimation.

Afterwards, a network between the points is generated, typically using a Delaunay triangulation. Other forms of connections, including multi-layered connections, are possible and can provide stronger results. The network of points is, in the following step, used to estimate the APS. For this, the network connections can have a maximal length. A typical maximum value is 2 km (Ferretti *et al.*, 2001) or below, which will be explained in the following section.

It is not always possible, though, to form a network with strict maximum edge length requirements. For example, with the Yangtze River separating Wuhan, it can be difficult to establish a network covering the whole city, as the width of the Yangtze is about 1 km to 2 km throughout the city. The connections crossing the river are few and only with relatively low temporal coherence, which can lead to network and APS estimation problems in the following processing. Parks, forests, mountainous areas, lakes, etc., can all lead to similar problems, but large rivers can be especially troublesome, as they can separate the network into two, or more, unconnected sub-networks.

6.4   Estimation of the Atmospheric Phase Screen

SAR interferometry is by its nature a relative measurement. That is to say, it does not give us absolute results, as we do not know the absolute phase value, but relative results, as phase differences between two points. Furthermore, the wrapping of the phase between $-\pi$ to π complicates the measurement immensely.

The next step in estimating the atmosphere is to derive the phase differences along the edges of the newly formed network. As each point in the network has a vector of complex values containing the values from all interferograms (flattened and DEM phase contributions removed) of the InSAR stack, the phase differences are formed by a complex conjugate multiplication of the corresponding values in each vector. Each edge now has an implicit direction, given by the order of the subtraction from the two points forming the edge, and a vector of differential phases attached to it. From these phase vectors, the parameters of the PSI process are estimated along the respective edges. Each differential phase value φ in a vector includes information on

$$\varphi = W\{\phi_{\text{topo}} + \phi_{\text{motion}} + \phi_{\text{atmo}} + \phi_{\text{orbit}} + \phi_{\text{noise}}\} \qquad (6.2)$$

where φ is the wrapped phase and $W\{\}$ is the wrapping operator. ϕ_{topo}, ϕ_{motion}, etc. are the separated contributions from the topography, motion, etc. The main idea in the PSI process is that these terms can be separated, as they behave differently in space and in time. ϕ_{noise} is typically modeled as random white noise and is therefore difficult to estimate and remove. However, the effects of the thermal noise are minimal, especially in the case of high SNR targets used in PSI. So, in the following, we will ignore this term. The error caused by the wrong estimation of the orbit ϕ_{orbit} is typically visible as a phase ramp and relatively easy to determine and remove. However, due to its spatial and temporal similarity to ϕ_{atmo}, orbit errors are often estimated together with the atmosphere. This will become easier to understand at the end of this section. So, for the moment, we also ignore this term or consider it as included in ϕ_{atmo}, so that

$$\varphi = W\{\phi_{topo} + \phi_{motion} + \phi_{atmo}\} \tag{6.3}$$

where ϕ_{topo} and ϕ_{motion} can now be estimated, for example, via a least-squares estimation using the available vector of phase values. ϕ_{res} is then the residual of

$$\phi_{res} = \phi - \phi_{topo} - \phi_{motion} \tag{6.4}$$

and ϕ_{atmo} is estimated based on ϕ_{res}. For this to work, ϕ_{res} has to be smaller than 2π, as the values are still wrapped. Actually, ϕ_{res} has to be much smaller than 2π. This is the reason for the size limitations on the networks. As we assume that ϕ_{res} is influenced mainly by the atmosphere and that atmospheric phase delay changes slowly over the image, then this assumption is true for short edges. The value of 2 km maximal edge length is a rule of thumb and we prefer even smaller edge lengths, especially under turbulent weather conditions.

6.4.1 Estimating Topographic Height Error and Velocity

From Chapter 3, we know that

$$\Delta h = \frac{\lambda \Delta \phi_{topo}}{4\pi} \frac{r \cdot \sin \theta}{B_\perp} \tag{6.5}$$

Similarly, the motion phase ϕ_{motion} can be estimated, as shown in Chapter 5:

$$\phi_{motion} = \frac{4\pi}{\lambda} \Delta r \tag{6.6}$$

To estimate the motion via a least-squares solution, we need to define a model for Δr. In the simplest case, we can estimate a linear motion, so that

$$\Delta r = \Delta v_{\text{linear}} \cdot \Delta t \tag{6.7}$$

where Δv_{linear} is the linear motion velocity to be estimated and Δt is the time difference, considered as the time difference between the master image and the slave image of each interferogram. We can then formulate

$$\Delta v_{\text{linear}} = \frac{\phi_{\text{motion}} \cdot \lambda}{4\pi \cdot \Delta t} \tag{6.8}$$

Having a vector of ϕ_{topo} values, with r, $\sin\theta$, and a known vector of $B_\perp$, the single height value Δh could be estimated as this is an overdetermined system. The only problem is that in fact it is not an overdetermined system, but an underdetermined one.

The problem again is the wrapping. We have to include vectors of unknown a_1 to a_n representing the wrapping factor for the topographic phase error component and b_1 to b_n for the motion phase component, where each a and b is an integer value, so that

$$\begin{bmatrix} \phi_1 \\ \vdots \\ \phi_n \end{bmatrix} = \begin{bmatrix} a_1 \cdot 2\pi + \phi_{\text{topo},1} + b_1 \cdot 2\pi + \phi_{\text{motion},1} + \phi_{\text{res},1} \\ \vdots \\ a_n \cdot 2\pi + \phi_{\text{topo},n} + b_n \cdot 2\pi + \phi_{\text{motion},n} + \phi_{\text{res},n} \end{bmatrix} \tag{6.9}$$

where $\phi_{\text{topo},1}$ is the wrapped topographic phase error of the first interferogram, $\phi_{\text{motion},n}$ is the motion phase component of the nth interferogram, and ϕ_n is the unwrapped phase of the nth interferogram, where n is the number of interferograms in the stack. If we consider the phase wrapping, which is necessary, the system is in fact an underdetermined systems, as a_1 to a_n and b_1 to b_n are unknown.

This process is sometimes also referred to as spatio-temporal unwrapping because solving these equations can be seen as an unwrapping in space (e.g. along the edges) and in time (e.g. from a_1 to a_n). The strategy used for solving this is a distinguishing factor between the different PSI approaches.

In the traditional PSInSAR approach, the idea is to minimize ϕ_{res} by estimating the values for Δh and Δv_{linear} that best fit to the observed ϕ.

In the PSInSAR approach, a periodogram with an irregular sampling of the two dimensions, baselines and time, is used to maximize the absolute temporal coherence $|\hat{\gamma}|$. For more information on the temporal coherence, you can refer to the following section; for the moment, we can understand that maximizing the temporal coherence is similar to minimizing the phase residual:

$$\underset{\Delta h, v_{\text{linear}}}{\arg\max} \left\{ |\hat{\gamma}| = \left| \frac{1}{N} \sum_{n=1}^{N} e^{j\phi} \cdot e^{-j(\phi_{\text{topo}} + \phi_{\text{motion}})} \right| \right\} \qquad (6.10)$$

Using the periodogram over the two dimensions allows for an estimation of the best fitting values for Δh and Δv_{linear}. This is not the only possible solution though. Another approach is the least-squares approach to solve for the solution with the least amount of residuals. Even more so, as we know that a and b are integers, an integer least-squares solution can be used, as introduced by Kampes (2006) using LAMBDA.

Furthermore, more complex models for Δr can be developed and integrated into the solutions considering that with the number of variables to be estimated, the number of interferograms should be increased accordingly to allow for an unambiguous solution.

Using a suitable solver, for example, the integer least squares, the values of Δh and Δv_{linear} can be estimated for each edge in the network. Furthermore, based on the estimated Δh and Δv_{linear}, ϕ_{topo} and ϕ_{motion} can be calculated for each interferogram and removed from the measured phase ϕ, so that a vector of residuals $\phi_{\text{res}, 1}$ to $\phi_{\text{res}, n}$ is available for each edge along the network.

6.4.2   Unwrapping along the Edges

Now, we have estimated height and velocity values as well as a vector of residual phase values along the edges. This is shown in Figure 6.2 exemplarily illustrating height values. To move from values along the edges to values for each PS point along the network, it is necessary to first

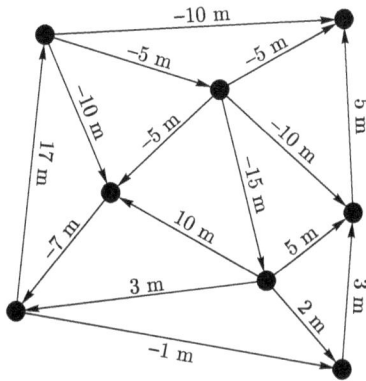

Figure 6.2 Example of height values estimated along the network edges.

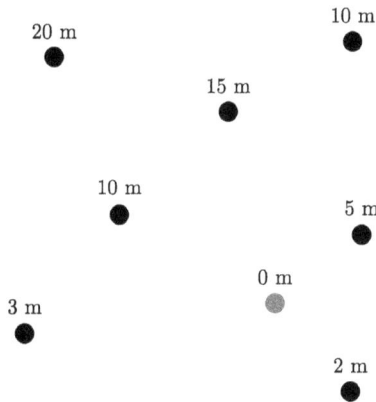

Figure 6.3 Derived height values for PS network points.

define a reference point. We define this point as zero height and having zero velocity. All further measurements are in relation to this reference point. It would also be possible to give previously known values of height or deformation, for example, from ground measurements, as starting values of the reference point. Going from the reference point, the points along the network are assigned values following the edges, as shown in Figure 6.3, where the gray point is the selected reference point.

For each point along the network, values for height and velocity are determined. Furthermore, the residuals are defined and also unwrapped along the network. The whole process can be interpreted as an unwrapping process along the edges for the residual phases as well as for the height and velocity values. In the standard PSI approach, the now unwrapped residuals at the network points are used to estimate the APS over the image.

Another point that has not been discussed is errors along the edges. In the example, we see perfectly estimated heights along the edges that all fit very well. In reality, errors occur though, be it from unwrapping errors/ambiguities or non-stable points used as network points, noise, etc. Errors happen! If two paths along the network lead to different results for a point, which one should be selected? This is not easy to answer, as different solutions exist. Certainly, one would suggest the most probable solution. This can be defined as the one having the lowest residuals, the shortest path, the shortest weighted path, etc. After the values of the points in the network are defined, each triangle can also have a sanity check, following the values along the triangle and looking for residuals. These might be repaired by including information from the neighboring triangles. A multi-layer approach can also assist here, where triangles are built on multi-scale layers that can help solve problems stemming from such inconsistencies.

The traditional PSI approach is less sensitive to such inconsistencies though, as only the residuals along the network are used and they are afterwards strongly filtered in space or time, so that small errors and irregularities in the network are mitigated. Other PSI solutions that do not estimate the APS but densify the network are much more sensitive to initial errors along the network in the estimation and therefore need better strategies for avoiding or repairing such problems in the network. See, for example, STUN in Section 7.1 or PSP in Section 7.2 of Chapter 2.

6.4.3   APS Estimation

For each point in the network, an unwrapped residual phase value for each interferogram is now available. We assume that the residuals contain information from the atmospheric phase delay and also residuals from imprecise height estimations and deviations from the linear velocity assumptions. This leads to residuals in the motion related phase, orbit error and noise. An additional filter step is necessary to separate the APS phase component.

The assumption on the properties of the atmospheric phase delay is that the phase delay is related in space, but not in time. Thus, there is only a small shift in the atmospheric phase delay expected from one pixel to its neighboring pixel, with a slow, graduate change in the phase delay. Phase delay is expected to have no relation at all to time. This has to be understood in the context of spaceborne SAR sensors though, as we have days between acquisitions, so that we can assume that the weather changes significantly between acquisitions. If we would have SAR acquisitions on, for example, an hourly basis, we would expect the atmosphere to be also related in time.

Under this assumption of atmospheric phase delay being correlated in space but uncorrelated in time, the residuals are filtered with a large low-pass filter in the spatial domain and a high-pass filter in the temporal domain. That is to say, each value is filtered by a mean filter (or other low-pass filter) in space. Typically, a rather large filter size is used. In time, an edge filter is used to find the part of the phase residual related to the respective atmospheric delay at the time and removes the static phase components that may come from errors in the topographic estimation or errors in the linear velocity estimation at that position.

After this process, a filtered and unwrapped estimation of the phase contribution from the atmospheric phase delay for each pixel in each interferogram is available.

6.5   Velocity and Residual Height Estimation for All PS

After removing the phase contribution from the atmospheric path delay, the remaining phase components are

$$\phi = \phi_{\text{topo}} + \phi_{\text{motion}} + \phi_{\text{noise}} \qquad (6.11)$$

We can assume that ϕ_{orbit} is also removed as the phase related to orbit errors shows a similar behavior as atmospheric phase delay: correlated in space, but uncorrelated in time. Otherwise, orbit-induced errors show a clear linear phase ramp and can be estimated and separated with relative ease. This might be necessary as orbit errors can be correlated in time for some sensors. Ignoring ϕ_{noise}, the remaining phase contributions from ϕ_{topo} and ϕ_{motion} can now be estimated in a way similar to the estimation, as described for along the network edges. However, in this instance the values are not estimated along the edges. After removing the contribution from the atmosphere, there is no limitation on distance when estimating the differences in residual height and velocity.

All PS points are analyzed with respect to the the same reference point as before. Therefore, from the phases of all PS (after removing the estimated ϕ_{atmo}), the phases of the reference point are subtracted, so that the estimated residual heights and motion parameters are with respect to this reference point. Now, similar to the estimation along the edges in APS estimation, the residual height and velocity parameters are estimated. We can call this an unwrapping in time in contrast to the unwrapping in space described, as discussed in Section 3.2 of Chapter 3.

Finally, the parameters for the residual height and the deformation model are estimated, and each point is associated with a residual height value, deformation value(s), and a vector for residuals. These residuals can then be used to estimate temporal coherence (see Section 6.7 of Chapter 6). Furthermore, assuming that the residuals are caused by motion elements deviating from the linear velocity, these deviations can be calculated and shown, providing a more precise view of the deformation. However, these are still based on the linear velocity model and deviate around this basic, and maybe false, assumption.

6.6   Including Alternative Motion Models

The standard PSInSAR approach allows for the inclusion of different solutions for the estimation of height and velocity. As the height, or more precisely, the residual height difference between the PS, is clearly defined, motion can have different causes and behavioral patterns. The basic assumption of a linear deformation pattern is often not a good model; however, it can be replaced and extended by any conceivable model that can form a relation of the variables describing the model with ϕ_{motion}.

It is easy to replace linear estimation with a step-wise linear model for estimating motion behaviors during different periods. Non-linear motion models can also be integrated. A common example for non-linear motion is the thermal expansion. Objects change shape, form, area, and volume with changing temperatures. The amount of thermal expansion depends on temperature differences and material properties. This expansion can be described by the thermal expansion coefficient α_L, where L indicates the linear expansion. In most cases, this can be expressed as

$$\frac{\Delta L}{L} = \alpha_L \cdot \Delta T \tag{6.12}$$

where L is the length of the object and ΔT is the temperature difference in Kelvin. Specializing on the thermal expansion of high-rise buildings, the difference in L can be estimated from the difference in residual height. In more general terms, including also horizontal expansion of, e.g. bridges, α_L and L can be merged and estimated together in PSI, especially if we are not actually interested in estimation of the true α_L, but only the overall temperature related motion. We can formulate

$$\Delta L_{\text{los}} = \alpha_{\text{los}} \cdot \Delta T \tag{6.13}$$

where ΔL_{los} describes the motion in the LOS direction and α_{los} is the thermal expansion coefficient of a PS in the LOS direction:

$$\phi_{\text{thermal}} = \frac{4\pi}{\lambda}\alpha_{\text{los}} \cdot \Delta T \tag{6.14}$$

with the overall motion than being

$$\phi_{\text{motion}} = \phi_{\text{linear}} + \phi_{\text{thermal}} \tag{6.15}$$

Δv_{linear} and α_{los} can then be jointly estimated as long as ΔT is known. ΔT can be derived from publicly available weather information. Similarly, additional linear or non-linear motion elements can be estimated. However, by increasing the variables to be estimated, the amount of equations should also be increased, which in this case means that more images are required for a stable estimation.

6.7   Temporal Coherence

The temporal coherence, or ensemble coherence, is a measurement of the fit of an estimated phase model and the real observations. High temporal coherence therefore shows a good fit between the model and the measured data. If the phase residuals after the parameter estimation are low, the temporal coherence will be high and vice versa. It can be calculated via

---

6.1. Temporal coherence:

$$\hat{\gamma} = \frac{1}{K} \sum_{k=1}^{K} \exp(je^k) \qquad (6.16)$$

---

where j is the imaginary unit and e^k is the difference between the observed and modeled phase in interferogram k. Following Kampes (2006), we denote the temporal coherence $\hat{\gamma}$ with a 'hat' to show that this is an estimate of the coherence.

The temporal coherence offers a filter possibility, allowing the reduction of noisy points in post-processing. This is a very useful feature, to some degree, that goes so far that all points are considered PS candidates and finally only the points with a high temporal coherence get 'promoted' to PS points. This viewpoint makes sense, as the PSC are selected only based on their amplitude dispersion, which is only an estimation of their phase stability. High temporal coherence is then a clear indication that a PSC is truly a PS. However, a low temporal coherence is not necessary an indication of the opposite. A PS may have low temporal coherence if the deformation pattern does not follow the assumed model even if all the measured phases are stable. This is, for example, the case in suddenly activated sinkholes or landslides. They may show linear velocity patterns over long periods; after activation, they may suddenly increase in velocity, often non-linear, e.g. rainfall-related. This sudden increase in velocity is a dangerous sign and should, in an ideal world, trigger a warning. In standard PSI, this can be a deviation from the linear velocity model, therefore a loss of temporal coherence, and we may end up filtering out these relevant points.

The temporal coherence is used throughout the processing to determine the 'best' or 'most likely' solutions. It is also an important value in the post-processing, often used to filter for the best points. This is certainly a valid strategy, as PS points with low temporal coherence are often unstable, indicating that they are not PS. However, one needs to keep in mind that a low temporal coherence is just a sign that the estimated model or the estimated model parameters do not fit the measurement. This can be the case if the point itself is unstable, but it can also be the case if the basic model assumptions, e.g. the assumption of a linear velocity, does not describe the reality.

6.8   Example

In the following, we explore standard PSI processing with an example from Las Vegas. The TerraSAR-X high-resolution spotlight image stack consists of 19 images acquired from a descending orbit. We will use the identical stack of images for examples with STUN and StaMPS later on in Sections 7.1 and 7.3 of Chapter 7 to ensure a better comparability. In the examples shown here it is not the goal to present the best possible results, but, on the contrary, point out to problems and issues of different processing techniques and demonstrate shortcomings. Processing results from other approaches and software implementations will look different.

After collecting the stack of SAR images used for PSI processing, the next step is to identify the master image used for co-registration, which will also be the master image for the interferograms in the PS-based processing techniques. In our example, we selected the image acquired on 2010-09-08 as the master image (see Figure 6.4). We selected it to ensure small temporal and spatial baselines. It is centered on the spatio-temporal baselines and therefore generates this 'star' formation in the spatio-temporal baseline distribution, as shown in Figure 6.5.

All slave images are afterwards resampled toward the coordinate system of the master image. In our example, the standard coarse-to-fine strategy is used. However, in this case, the co-registration and resampling of the slave images needed to be checked carefully, as we found that two out of the 19 images were not co-registered correctly on the first attempt. In this example, that seemed to be due to the changing structures of the parked cars in the parking lots outside of the Las Vegas Convention Center and the various hotels and casinos in the image. The changing patterns of the cars seemed to influence a high enough percentage of the 1200 image chips that we selected for the estimation of the resampling parameters, so that two images were not resampled correctly.

Figure 6.4   TerraSAR-X high-resolution spotlight amplitude image of Las Vegas, acquired on 2010-09-08 (© DLR, 2010).

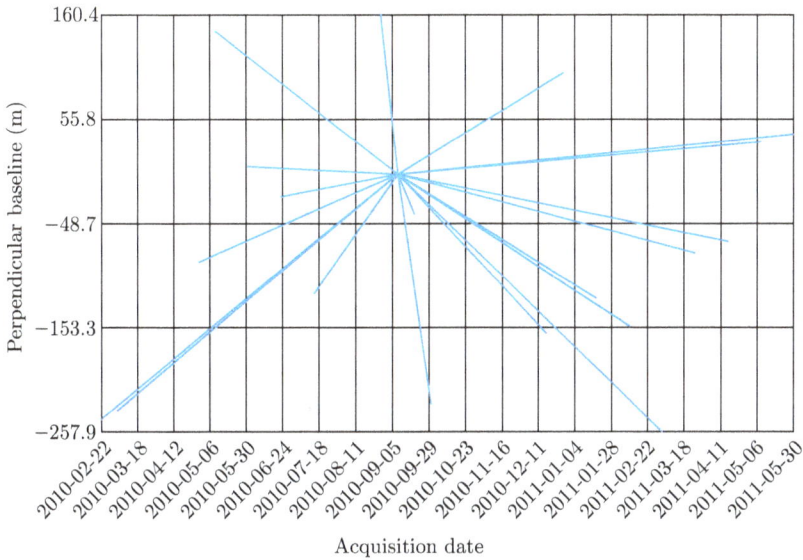

Figure 6.5 Spatio-temporal baseline distribution of the TerraSAR-X high-resolution spotlight data stack over Las Vegas used in this example.

Resampling errors can happen and because of this, the resampling results need to be checked carefully. Errors in this early stage can lead to serious problems in the velocity estimation later on, but are hard to detect in later steps. By selecting a larger number of image chips, 2800 in this case, the mis-registered images have also been resampled correctly.

After resampling, the PS candidates (PSC) are selected. We selected points with an amplitude dispersion index below 0.25. In this example, 2 990 534 points were selected. For these points, the interferometric phase after removal of the DEM is calculated. We used SRTM v4 in this example.

For estimating the atmospheric phase screen, we select only a subset of the PSC. The selection of the subset is based on selecting the best available PSC while ensuring a good spatial distribution. In our example, we divided the image in cells of 50 m size and selected the point with the lowest amplitude dispersion index in each cell. Afterwards, the points have been connected using a Delaunay triangulation. Edges exceeding

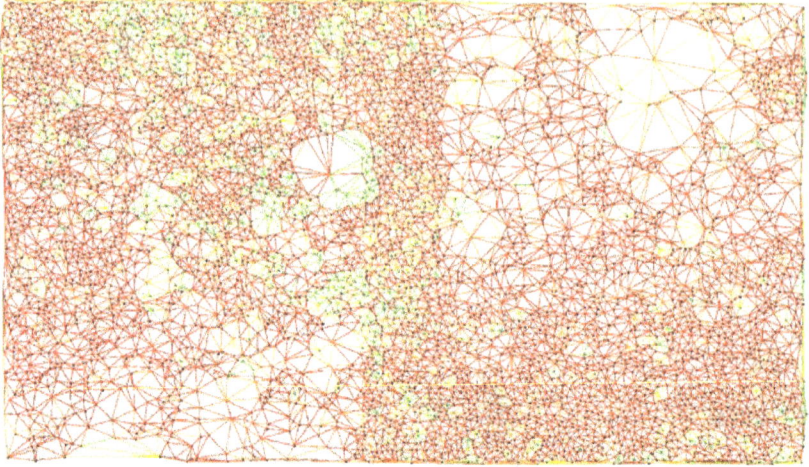

Figure 6.6   PSC connections and their estimated temporal coherence.

a 500 m distance have afterwards been discarded. 19 381 edges are afterwards left for processing (see also Figure 6.6).

For each connection, the differential velocity and topographic height error are calculated, and the temporal coherence and a vector of residuals are stored for each of the connecting vectors. In the next step, a reference point is defined. The reference point should be stable and ideally be located more toward the center of the image. Starting from this reference point, with an estimated topographic height error and linear velocity of zero, the vectors are traced toward each of the network nodes used for the APS estimation, summing the velocities and height estimations along the edges as well as phase unwrapping the residuals following the edges. This can be understood as an unwrapping along these edges. Finally, a height error estimation, a linear velocity as well as unwrapped residuals are available for each of the subset of PSCs.

The APS is now estimated based on these residual vectors because it is assumed that the residuals for each point are dominated by atmospheric effects. Nevertheless, the residuals are strongly filtered under the assumption that the APS is correlated in space but uncorrelated in time. Accordingly, a low-pass filter is used in space and a high-pass filter in time. The low-pass filter we use here has a dimension of 1 km.

After this operation an APS of 50 m cell size is estimated for each interferogram. Two examples are shown in Figure 6.7. As the overall dimension of the high-resolution spotlight images is limited to 10 km × 5 km, the APS images are rather small in this case.

This estimated APS is then removed from the phases of each PSC. Now, we consider the atmosphere to be removed. Without the atmosphere, there is no need for spatial proximity of the estimations. In the final step, each velocity and topographic height difference is directly estimated with respect to the reference point. For each of the PSC, a connection is formed from the reference point and the velocity and topographic height error difference with the reference point is estimated. As the atmosphere is removed, this can also be done for points at a large distance to the reference point. The results of this process are shown in Figures 6.7 and 6.8, containing 2 271 934 PS with a temporal coherence of the final estimation above 0.8.

The velocity results in Figure 6.8 show the subsidence center near the Las Vegas Convention Center. Some motion is also estimated along high-rise buildings, these however are residuals of temperature dilation of these buildings, a motion that has not been estimated in our example and that here is falsely interpreted as linear velocity. Finally, we can see an estimated uplift on the left edge of the image. This is an error. We would assume it comes from a wrong estimation of the APS, but the error is already in the estimated network, as we can see later on in the example demonstrating the STUN approach in Section 7.1 of Chapter 7.

(a)   (b)

Figure 6.7   Estimated APS over Las Vegas on 2010-12-16 (a) and 2011-01-18 (b).

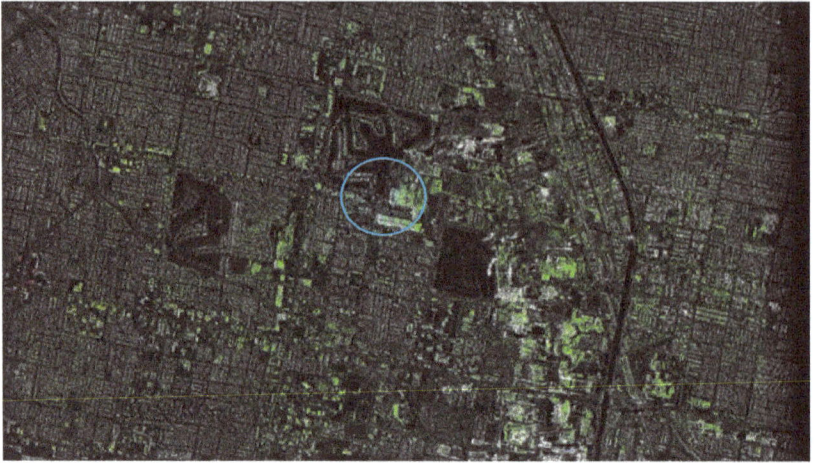

Figure 6.8   Estimated PS velocity over Las Vegas.

Figure 6.9   Estimated PS residual height over Las Vegas.

The estimation of the topographic height error in Figure 6.9 shows the relatively good estimation of the heights. The majority of points are estimated correctly and we can see some correct height estimation among the high-rise buildings.

Finally, Figure 6.10 shows how the estimation of the velocity would look without removing the atmosphere. In this example, we only performed the last step of the operation, estimating the height and velocity differences between each point and the reference point, but without removing the APS phase components estimated before. Although the area of interest is very small in this example and although Las Vegas is a desert area with relatively limited APS variety, the velocity estimations in Figure 6.10 are unusable. There is a difference in the visualization of the normal velocity results in Figure 6.9 and the results without APS correction in Figure 6.10, though. In the example without APS correction in Figure 6.10, there was no threshold used, as the temporal coherence of the points without correcting the atmosphere was mostly far below the threshold of 0.8.

Figure 6.10   Estimated PS velocity over Las Vegas without APS estimation and correction.

Chapter 7

Alternative Approaches to PSI

After the presentation of the PSInSAR approach by Ferretti *et al.* (2001), it became evident that this method reveals linear motion in urban areas. In areas with sparse PS density or non-linear motion, it does not work as well. The basic idea of PSI is the separation of the different phase contributions. They can be modeled and estimated using a large number of SAR images. Various other approaches have been developed over time that can sometimes provide better results. No approach works best under all circumstances, so a processing approach should be selected based on the area, expected deformation patterns, and available data.

7.1 Spatio-Temporal Unwrapping Network

The Spatio-Temporal Unwrapping Network (STUN) developed by Kampes (2006) is in many regards quite similar to the PSInSAR approach.

7.1.1 Point Selection in STUN

The point selection method of STUN is based on the average SCR of a pixel being above a threshold (Kampes, 2006). Based on Adam *et al.* (2004), the relation of the SCR to the phase error is

$$\sigma_\phi = \frac{1}{\sqrt{2 \cdot \text{SCR}}} \qquad (7.1)$$

An SCR threshold of 2 leads to a phase standard deviation $\sigma_\phi = 0.5$ rad (approximately 30°). As the SCR of a target in a resolution cell is unknown, it is typically estimated by calculating the average pixel values in the neighborhood and assuming that the background of the pixel in question is similar (see Figure 7.1).

This method works well in areas with a uniform background, but is more difficult to apply, for example, in urban areas. Several pixels not representative of the background of the resolution cell are included in the background calculation, when using this approach in dense urban areas. The SCR of the point-like scatterers themselves will be underestimated. The threshold in urban areas is, therefore, set relatively low.

The advantage of the SCR-based approach is that the amplitude data do not need to be calibrated, and no assumption on the temporal

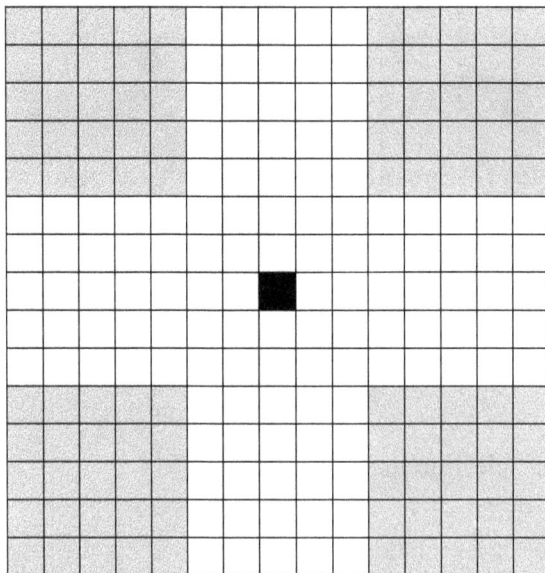

Figure 7.1   Calculating of the SCR: estimating the background values from the neighborhood pixels in gray.

amplitude behavior of the pixel is necessary (Kampes, 2006). The typically used amplitude dispersion index does not require estimation of the neighborhood information, but at least 20 images are needed for a reliable estimation of D_A (Adam *et al.*, 2004).

For high SCR, the estimation of the phase variance using the amplitude dispersion and the SCR method are equivalent. The SCR method, however, has a smaller bias compared to the amplitude dispersion index-based approach (Adam *et al.*, 2004).

7.1.2   Sparse Network Creation

Similar to PSInSAR, a reference point is required. The topographic residuals and velocity of this point are estimated to be zero. Thus, the values are estimated relatively along the network, so that the height and velocity differences along the edges of the network are unwrapped and for each network point, height and velocity values relative to the reference point are determined.

7.1.3   Solving the Equations with LAMBDA

One big difference between STUN and PSInSAR is the way they solve the equations for estimating the height difference, velocity, and residuals. In the traditional approach, a direct search of the solution space is used, maximizing the temporal coherence via a periodogram approach. STUN uses the Least-squares AMBiguity Decorrelation Adjustment (LAMBDA) method instead. LAMBDA was developed for fast GPS double-difference integer ambiguity estimation (Teunissen, 1995). As described in Chapter 6, we know that the unwrapped phase is an integer multiplied by 2π, therefore an integer least-squares approach can be used. The requirements for the integer ambiguity solution in GPS and PSI are rather similar, so that using LAMBDA is rather straightforward.

In PSI, the problem is always underdetermined, as for each observation, the phase ambiguity is unknown.

Similar to PSInSAR, a reference point is required. The topographic residuals and velocity of this point are estimated to be zero. Then the values are estimated relatively along the network, so that the height and velocity differences along the edges of the network are unwrapped, and for each network point, a height and velocity value relative to the reference point are determined.

7.1.4   Densifying the Network

In the PSInSAR approach, the residuals of the phases after estimating the height and velocity values are used to estimate the APS and then removed. In STUN, the APS is not estimated, instead the sparse network is densified using the estimated values of the sparse network points.

The basic idea for setting up the network is that along short edges, atmospheric influence is minimal, so we can estimate height and velocity while keeping the influence of the APS and noise far below 2π. The same is true for points close to nodes on such a network. After unwrapping along the edges, the points can be densified by calculating height and velocity differences to the closest node in the unwrapped network. In this way, the actual removal of the APS is not necessary. This is unlike the PSInSAR method, where the estimated APS is removed and the difference in phase and velocity to each point from a selected reference point is calculated.

The advantage of STUN therefore lies in skipping APS estimation altogether and relying on the unwrapping along the network. This works well as long as the estimation along the edges is correct. In PSInSAR, errors in the estimation along the edges can, at least partly, be corrected by the large spatio-temporal filter used to estimate the APS. The large filter used in PSInSAR, however, can also lead to errors in the unwrapping of the atmospheric phase.

7.1.5 Example

The same data stack as in Chapter 6 after pre-processing was used for comparison. The 2 990 534 points were initially selected based on the amplitude dispersion index. From these points, a subset was connected to 19 381 edges, as in the example above.

This example does not exactly follow the STUN processing, as the amplitude dispersion index has been used for PSC selection instead of the SCR-based approach. This ensures the maximum comparability between the results.

After the estimation of the values along the edges, the information on the edges is unwrapped, starting from a reference point to all the nodes in the network. In the standard PSInSAR processing, the residuals in the nodes are afterwards used to estimate the APS, while the rest of the information is not used anymore.

In STUN, there is no estimation of the APS though. After estimating the values for the linear velocity and topographic error for all nodes in the initial network, the values are then estimated for all of the 2 990 534 points by estimating the closest network node for each point and estimating the differences between the already unwrapped values of the closest node and the point in question.

As each of these connections is then again short enough to be not strongly influenced by the APS, the correct estimation of the APS is not necessary. Finally, 2 856 397 PS points with a temporal coherence above 0.8 are found, as shown in Figures 7.2 and 7.3. The number of PS is slightly higher than that in the PSInSAR approach, however the temporal coherence is in this case only with relation to the closest node point.

From Figure 7.2, we can see that the estimation of the velocities is generally good, while the estimation of the topographic error worked very well, as shown in Figure 7.3. The errors in the estimation are larger though in this result compared to the PSInSAR result. The large outlier on the left side of Figure 7.2 is stronger than in the PSInSAR result in Figure 7.2. The error is therefore in the estimation of the values along the network edges and slightly mitigated by the APS filtering in the PSInSAR process, whereas such network errors remain in STUN.

Figure 7.2   Estimated velocities of the Las Vegas stack using an STUN-based approach.

Figure 7.3   Estimated residual heights of the Las Vegas stack using an STUN-based approach.

Better network estimation, e.g. by multi-layered approaches, is therefore important in STUN or STUN-like approaches like the PSP approach shown in the following section.

7.2   Persistent Scatterer Pair Interferometry

A recent approach to PSI that is related to the STUN approach is the Persistent Scatterer Pair Interferometry (PSP) (Costantini *et al.*, 2014). PSP is, similar to STUN, built around the network of PS points. In the PSInSAR approach, the network of PS is only used to determine the APS, which is then subsequently removed, so that the estimations for topographical residuals and relative motion are calculated with respect to the selected reference point for each PS. PSP relies on the network completely, making the explicit estimation of the APS unnecessary.

In contrast to STUN, PSP includes all PS points in the network from the beginning. This specific difference probably is related to the increase in processing power seen over the decade of development in information technology that lies between the two approaches, allowing for the inclusion of many more points in the network. Unlike STUN, PSP solves estimation along the arcs using a periodogram approach, staying closer to the traditional PSInSAR approach in this regard.

With all PS included in the network, the values are estimated along the arcs. As each arc is limited in its distance, atmospheric effects, as well as other error sources like orbit errors, are minimal and do not significantly disturb the estimation along the arcs. The propagation of the values along the arcs can be understood as an unwrapping along the network.

However, in PSP, great care has to be taken to avoid error propagation along the network as it may directly influence the final estimations. In the traditional PSInSAR approach, error propagation along the network can be, at least partly, mitigated by the later filtering in space when estimating the APS.

Similar to 2D phase unwrapping, having a dense network between the PS permits a search for residuals and errors in the propagation of the values along the network, thus improving the estimations of the values. Multi-layered network approaches in the arcs can further help solve ambiguities and avoid error propagation in the network.

7.3   Stanford Method of Persistent Scatterers

The Stanford Method of Persistent Scatterers (StaMPS) was proposed by Hooper *et al.* (2004). Due to its stability, it is probably the most widely used PSI method, especially because the software is freely available. StaMPS was developed for use in geophysical applications targeting, in particular, the measurement of volcanic surface motion. Volcanoes have a reduced amount of PS, although exposed rocks and sparse vegetation on many volcanoes provide some stable points. The PS density is therefore not the main issue.

Volcanoes, however, do not show linear motion; moreover, there is no applicable model for volcanic deformation. The spatio-temporal deformation patterns are important to learn more about volcanic activity. StaMPS therefore does not require a deformation model and can estimate linear and non-linear motion.

To achieve that goal, the residual height is estimated, based on the assumption that this is the only phase component related to baseline differences. The velocity subsequently is estimated by a 3D phase unwrapping process, but requires the estimated surface motion to be related in space. Sudden "jumps" in the surface motion can cause unwrapping problems and an underestimation of the estimated motion. Volcanic motion, however, is typically related in space and does show a gradual increase, making the basic assumptions of StaMPS feasible whereas motion patterns in urban areas or landslides may show sharp spatial edges, making StaMPS less suitable.

7.3.1   Persistent Scatterer Selection in StaMPS

PS are defined by their phase stability, but the previously presented approaches, such as the Amplitude Dispersion Index as well as the SCR-based methods, rely on amplitude as a proxy for phase stability. This works quite well in both methods for points with strong backscattering, but does not work well for phase-stable points with low SCR. Exposed rocks and other natural PS are common on volcanoes.

They can remain stable and can be suitable for interferometric measurements, even with comparably low backscattering strength. So, to achieve a higher PS density, StaMPS must find such points. As a certain signal-to-clutter ratio is necessary for interferometric measurements, StaMPS typically uses the Amplitude Dispersion Index for a first point selection, but with a much less restrictive threshold of 0.4. If necessary, this threshold can be increased, but the likelihood of finding suitable points is drastically reduced at higher thresholds. Meanwhile, the computational complexity increases with higher thresholds. A threshold of about 0.4 is, therefore, suitable in most cases.

So, comparable to the phase model presented previously, we can separate the phase contributions to

$$\phi = \phi_{\text{topo}} + \phi_{\text{motion}} + \phi_{\text{atmo}} + \phi_{\text{orbit}} + \phi_{\text{noise}} \tag{7.2}$$

While processing each PSI approach, ϕ_{noise} should be small, so that it does not significantly disturb the measurement. In StaMPS, the PS are defined as points with very low ϕ_{noise}.

In StaMPS, it is assumed that ϕ_{motion}, ϕ_{atmo}, and ϕ_{orbit} are spatially correlated over a given length l and ϕ_{topo} and ϕ_{noise} are uncorrelated over the same distance. This is quite similar to the assumption in PSInSAR for estimating the APS, but differs in that ϕ_{motion} is also assumed to be correlated in space. This assumption is often true for motions in applications like volcanology or tectonics, but might not be true for small-scale phenomena like sink-holes or landslides that may not show spatial correlation.

If all PS are known, averaging the phase of all PS within a circle centered on the analyzed pixel with the radius of l, we get

$$\bar{\phi} = \bar{\phi}_{\text{motion}} + \bar{\phi}_{\text{atmo}} + \bar{\phi}_{\text{orbit}} + \bar{\phi}_{\text{noise}} \tag{7.3}$$

The bar denotes the mean of the sample and $\bar{\phi}_{\text{noise}}$ is the mean of $\phi_{\text{noise}} + \phi_{\text{topo}}$. Subtracting these two equations, we get

$$\phi - \bar{\phi} = \phi_{\text{topo}} + \phi_{\text{noise}} - \bar{\phi}'_{\text{noise}} \tag{7.4}$$

where $\bar{\phi}'_{\text{noise}}$ is the sum of $\bar{\phi}_{\text{noise}}$ plus the difference between the patch mean and the values for the deformation, atmosphere, and orbit errors. It is therefore comparable to the residuals along the edges in the PSInSAR and the STUN approach.

ϕ_{topo} is the only part of the phase that is related to the perpendicular baseline $B_\perp$, which can be used to estimate ϕ_{topo} at this early stage already. In PSInSAR, ϕ_{topo} is estimated together with the velocity, but in StaMPS, the estimation of these values is separated. ϕ_{topo} depends directly on the perpendicular baseline, so that

$$\phi_{\text{topo}} = B_\perp \cdot k_\varepsilon \qquad (7.5)$$

where k_ε is a constant. k_ε can be estimated using all interferograms in a least-squares approach, and from k_ε, the residual height of each point can be derived. If we substitute the expression, we get

$$\phi - \bar{\phi} = B_\perp k_\varepsilon + \phi_{\text{noise}} - \bar{\phi}'_{\text{noise}} \qquad (7.6)$$

In StaMPS, the temporal coherence γ is defined as

$$\gamma = \frac{1}{N} \left| \sum_{i=1}^{N} \exp\{j(\phi - \bar{\phi} - \hat{\phi}_{\text{topo}})\} \right| \qquad (7.7)$$

where N is the number of interferograms and $\hat{\phi}_{\text{topo}}$ is the estimate for ϕ_{topo}. γ is a measure of the phase stability and therefore an indicator if the analyzed point is a PS. As StaMPS requires the calculation of the mean of the known PS, but at this stage in the processing, the PS are unknown, an iterative process is used. To select the PS from the PSC, a threshold for γ is needed. The goal is to maximize the number of real PS while minimizing the false positives.

After selecting the PS, the estimated $\hat{\phi}_{\text{topo}}$ is removed from the phase values. This is a necessary step before the upcoming three-dimensional phase unwrapping, as the topographic phase is not related in space and can lead to phase discontinuities during unwrapping. The remaining phase components in the StaMPS approach are either assumed to be correlated in space, like ϕ_{motion}, ϕ_{atmo}, and ϕ_{orbit}, or they are supposed to be small as in ϕ_{noise}. The estimation of k_ε however might be erroneous, hence the removal of the topographic phase may have a residual phase component, but this phase component is assumed to be small and spatially correlated.

7.3.2   3D Phase Unwrapping

Unwrapping in PSI can be understood as 3D phase unwrapping, with two dimensions in space and one dimension in time when a time series of interferograms is available. Moving from 1D phase unwrapping to 2D phase unwrapping simplifies the solution, so that more paths can be found. Similarly, 3D phase unwrapping is also beneficial. As it is important to use the advantages from 2D phase unwrapping over 1D phase unwrapping, by treating it as a true 2D problem instead of a series of 1D problems, it is also better to treat the 3D phase unwrapping as a single problem.

However, there is no 3D unwrapping algorithm that is suitable in terms of general applicability and efficiency. Thus, StaMPS implements a Pseudo-3D algorithm. The data are first unwrapped in one dimension, time, and this is used as an initial solution for optimizing the unwrapping in the two spatial dimensions. This has also another advantage because it allows the already existing and efficient 2D unwrapping algorithms to be used.

Compared to PSInSAR, arcs between the PS are formed, as the difference in phase of nearby PS is only minimally influenced by the APS. The phase differences between the arcs are unwrapped in time. Snaphu (Chen & Zebker, 2001) is used for the spatial unwrapping. This requires the data to be regularly gridded. Therefore, StaMPS uses an iterative weighted least-squares approach. The approach iterates, dropping arcs with the largest residuals until all residuals are zero.

7.3.3   Spatially Correlated Terms

After unwrapping for φ_{motion}, the remaining error terms, ϕ_{atmo}, ϕ_{orbit}, and ϕ_{noise}, are assumed to be spatially correlated but uncorrelated in time. Similar to the estimation of the APS in the PSInSAR approach, by using high-pass filtering on the unwrapped data in time and a low-pass filter in space, the spatially correlated error term can be estimated and then removed.

There is another remarkable difference between StaMPS and other PSI methods. Since the unwrapping is not along the network starting from a reference point. Therefore, instead of relating the velocity values to a single reference point, StaMPS indicates the motion relative to the motion of the average of the points. It is however also possible to relate the motion to an area or to a single point (Figure 7.4).

Figure 7.4   Estimated PS velocity over Las Vegas with a StaMPS-based approach.

Table 7.1 shows some common PSI methods.

Table 7.1 Overview of PSI and DS-InSAR methods (modified from Crosetto *et al.*, 2016; Liao *et al.*, 2020; Minh *et al.*, 2020; Osmanoglu *et al.*, 2016; Zhang *et al.*, 2012).

Technique	Baseline configuration	Point selection	Deformation Model	Year	Reference
PSInSAR	Single Master	Amplitude Dispersion	Linear Deformation	2001	Ferretti *et al.* (2000, 2001)
SBAS	Small Baselines	Spatial Coherence	Spatial Smoothness	2002	Berardino *et al.* (2002)
IPTA	Single Master	Amplitude Dispersion	Linear Deformation	2003	Werner *et al.* (2003)
StaMPS	Single Master	Phase Stability	Spatial Smoothness & 3D Unwrapping	2004	Hooper *et al.* (2004)
Multi-DIFSAR	Small Baselines	Spatial Coherence	Spatial Smoothness	2004	Lanari *et al.* (2004)
STUN	Single Master	Signal-to-Clutter Ratio	Linear Deformation	2006	Kampes (2006)
CPT	Small Baselines	Spatial Coherence	Conjugate Gradient Method	2008	Blanco *et al.* (2008)
SPN	Small Baselines	Spatial Coherence	Stepwise Linear Deformation	2008	Crosetto *et al.* (2008)
QPS	Coherent Baselines	Quasi-PS Approach	Linear Deformation	2011	Perissin and Wang (2011)
SqueeSAR	Full Graph	Statistical Homogeneity	Different Deformation Models	2011	Ferretti *et al.* (2011)
TCPInSAR	Small Baselines	Offset Deviation	Linear Deformation	2012	Zhang *et al.* (2012)
MInTS	Small Baselines	Spatial Coherence	Different Deformation Models	2012	Hetland *et al.* (2012)
PSP	Single Master	Amplitude Dispersion	Linear Deformation	2014	Costantini *et al.* (2014)
DSI	Small Baselines	Statistical Homogeneity	Linear Deformation	2014	Goel and Adam (2014)
Cousin PS	Small Baselines	Amplitude Dispersion	Spatial Smoothness	2014	Devanthéry *et al.* (2014)
Tomography Add-On	Single Master	Spectral Diversity	Different Models and Tomography	2016	Siddique *et al.* (2016)
Sequential Estimator	Efficient Stacking	Statistical Homogeneity	Linear Deformation	2017	Ansari *et al.* (2017)
Polarimetry-based PSI	Full Graph	Statistical Homogeneity	Linear Deformation and Polarization	2018	Mullissa *et al.* (2018)
Parallel SBAS	Small Baselines	Spatial Coherence	Spatial Smoothness	2019	Manunta *et al.* (2019)

Chapter 8

Distributed Scatterer Interferometry

Permanent scatterer interferometry (PSI) is a powerful toolset, overcoming the problems of atmospheric disturbances, speckling, and temporal decorrelation. Actually, the problem of temporal decorrelation and speckling are not overcome, but they are avoided by concentrating on the Permanent Scatterer (PS) that do not suffer from speckling and temporal decorrelation.

A PS is defined as having only one dominant scatterer in a resolution cell and therefore does not suffer from speckling. Furthermore, a PS is to be phase-stable over time, therefore not suffering from temporal decorrelation. However, PS is just a fraction of the pixels in an SAR image. They are mostly formed from man-made objects because man-made objects often form dihedral (walls) or trihedral (edges) structures that backscatter a large amount of energy and stay phase-stable over longer periods of time. PSI therefore works well in urban areas or areas with many artificial structures that form PS.

In non-urban areas, only a limited amount of PS are to be found. If there is no PS in the area of interest, no measurement is possible with PSI. Even worse, a certain density of points, more than two points per square kilometer, is necessary to have a measurement dense enough to successfully estimate the APS. Similarly, methods that do not precisely estimate the APS, still need a dense network of points, so that the length of the network edges stay below the critical distance.

If deformation measurements in areas with only sparse PS density is required, methods that do not (only) rely on PS points are necessary. To some degree, StaMPS (see Section 7.3 of Chapter 7) was already an approach for this problem by trying to include more PS. StaMPS not only looks for points with high amplitude but also searches more generally for phase-stable points. StaMPS was developed for applications in volcanology and so for non-urban areas. Volcanoes however show a relative high density of naturally occurring PS on, for example, exposed rocks. StaMPS also relies on PS, but accepts more points with lower amplitudes.

Methods that do not rely on PS alone are technically not PSI methods. Nevertheless, there are often more or less referred together, as in all of the methods, several aspects of PSI are still involved. These methods are often categorized as multi-baseline interferometry techniques, as all of them use a large number of interferograms. Recently, the term DS-InSAR is becoming a more commonly accepted term for these methods, where DS stands for distributed scatterer.

In dealing with DS, the speckle effect is disturbing the phase. To deal with this quasi-random effect, information from several pixels has to be combined to allow for a statistical analysis, at the very least forming an average over several pixels.

DS often also suffer more from temporal decorrelation as well as from geometrical decorrelation. DS-InSAR deals with these to some degree, but strong temporal, spatial, or volume decorrelation cannot be corrected with DS-InSAR. Furthermore, due to the more complex nature of DS, for example, the occasional dependence of the observable scattering center on soil or plant moisture, observed motions can be related to changes in moisture. Ansari *et al.* (2020) show the existence of a systematic bias in measuring surface deformation with SAR interferometry within short temporal baselines and using DS.

8.1   Small Baseline Subset Technique

The first DS-InSAR technique was the Small BAseline Subset (SBAS) technique (Berardino *et al.*, 2002). It was published shortly after the introduction of PSInSAR (Ferretti *et al.*, 2000, 2001). SBAS is also a very commonly used technique, as it is supported by several software packages, including the freely available StaMPS.

SBAS reduces the temporal and spatial decorrelation by forming the interferograms between the SAR images in a way that minimizes the temporal and spatial baselines. In the PSI approaches, the interferograms are formed between a single master image and all other slave images. Therefore, some interferograms will have rather large spatial or temporal baselines, which is not a huge problem when dealing with PS; PS are stable over time, so that large temporal baselines are not an issue and are relatively insensitive to spatial decorrelation. DS, on the other hand, are sensitive to both effects, but they can be mitigated by reducing the temporal and spatial baselines in the interferograms.

The differences in the combination of the interferograms can be seen in Figure 8.1. The name of the approach comes from the combination of interferograms with small baselines, specifically those with small temporal and small perpendicular baselines. A maximum for the temporal and the perpendicular baseline is defined, and then the interferograms fulfilling these requirements are formed. Normally, more interferograms are processed in SBAS than in the PSI approaches, making it more computationally intense.

Temporal and spatial decorrelation are mitigated by processing only interferograms with small baselines. Speckling is still a problem. SBAS is reducing the influence of speckling by multi-looking the interferograms. Multi-looking, a process of averaging over several pixels, reduces

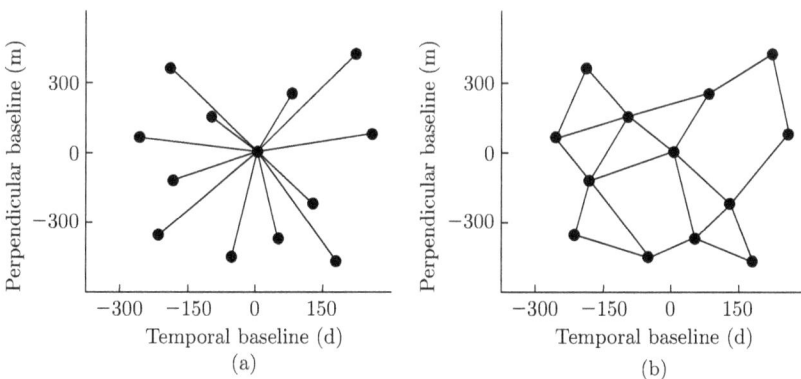

Figure 8.1 Combination of interferograms, with the temporal baseline in days on the *x*-axis and the perpendicular baseline on the *y*-axis. (a) star graph; (b) SBAS combination.

the effect of speckling (see Section 2.5.1 of Chapter 2). In PSI, the interferograms are not multi-looked, as PS are defined to be one dominant scatterer in a resolution cell, so there is no speckling effect to be reduced, and multi-looking may destroy the properties of the PS as multiple pixels are combined, possibly reducing the dominance of the PS.

After multi-looking and interferogram generation, SBAS unwraps the interferograms. SBAS uses 2D unwrapping for each interferogram, therefore the phases are first unwrapped in space. For large motion gradient differences, this can lead to unwrapping errors and underestimation of the motion, but as the temporal baselines are kept short, this problem is also reduced. Pixels with a spatial coherence below a certain threshold are exempt from the phase unwrapping and hence exempt further processing.

In the next step, the low-pass component of the deformation and possible topographic errors are estimated using a least-squares solution. These are removed from the phases. The interferograms can then be unwrapped again to reduce the fringe density and improve the unwrapping. SBAS then estimates a linear velocity using an SVD approach.

Atmospheric effects remain in the velocity estimation so far and need to be removed. This step is similar to the PSInSAR approach, as the atmospheric phase component is estimated using a low-pass filter in space and a high-pass filter in time. The contribution of the atmospheric effect to the velocity estimation is removed for the final velocity estimation.

8.2   Quasi-Persistent Scatter Technique

The Quasi-Persistent Scatterer (QPS) technique (Perissin & Wang, 2011) increases the number of targets in PSI by including targets that are only partially coherent, so PSI measurements in non-urban areas with sparse PS become possible. QPS uses approaches from SBAS and integrates those into the PSI framework. Similar to SBAS, the interferograms are

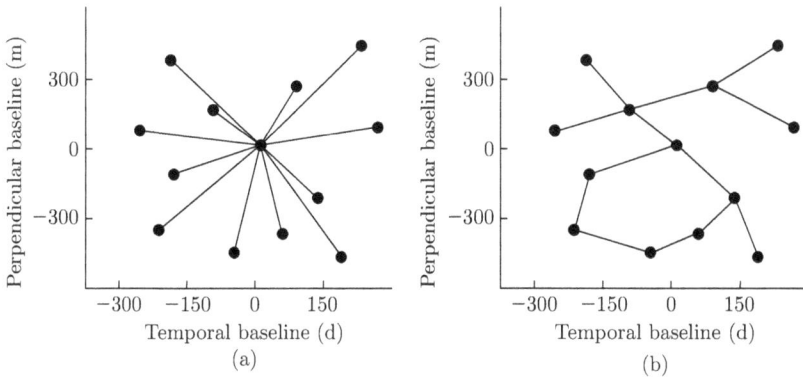

Figure 8.2 Combination of interferograms, with the temporal baseline in days on the *x*-axis and the perpendicular baseline on the *y*-axis. (a) star graph; (b) QPS combination.

not formed with respect to a single master image. The interferograms selected for the height and velocity estimation can be selected differently for different targets. Also, QPS uses spatial filtering of the interferograms to reduce speckling of DS.

When processing point-like targets, a star graph allows for a good distribution of temporal and perpendicular baselines and therefore good height and velocity estimation, including non-linear velocity estimation models. However, this is only for PS as they can keep coherent over time and are less affected by spatial decorrelation.

In QPS, a minimum spanning tree (MST) is built to connect the interferograms with the highest spatial coherence (Figure 8.2). The coherence of the combinations can therein be estimated on a model of the critical spatial and temporal baselines or by estimating the average spatial coherence on a subset of points for each interferogram.

Nevertheless, for each point in the scene, the optimal set of interferograms that carry information can be different. PS are, by definition, coherent in all interferograms, other targets can be only partially coherent, that is, they are coherent in some interferograms. QPS extends the PSInSAR approach by introducing a weight into the solution for the height and velocity parameters. Each interferometric phase is therein weighted according to the spatial coherence, ensuring that only coherent interferograms are included.

QPS is processed on filtered interferograms to reduce the speckling effect on the DS. This adds more noise to the PS points that are also processed and reduces the spatial resolution, but it is necessary for processing the DS. The precise filtering is not exactly defined in the QPS processing, but multi-looking and phase filtering are suggested.

QPS includes elements from DS InSAR, while keeping close to the PSI method. QPS is a predecessor of the SqueeSAR approach as it includes DS methods, but staying as close as possible to PSInSAR.

8.3   SqueeSAR

As mentioned, a disadvantage of PSI techniques is the sparsity of PS in non-urban areas. SqueeSAR addresses this issue by including DS into the PSInSAR processing scheme. The goal of SqueeSAR is to process PS and DS together without the need for significant changes in the traditional PSInSAR processing chain (Ferretti *et al.*, 2011).

SqueeSAR uses an adaptive filtering strategy to combine similar pixels of DS while keeping point-like scatterers unfiltered. Such a strategy better preserves the phase homogeneity of the DS and allows for a joint processing of PS and DS.

SqueeSAR uses the DespecKS algorithm for this. DespecKS is looking at statistically homogenous pixels (SHP). Two pixels of a data stack are statistically homogenous if the null hypothesis that the two vectors have the same probability distribution function cannot be disproved. In DespecKS, the Kolmogorov–Smirnov (KS) (Stephens, 1970) test is used for this, but there are also other statistical tests available, that could be used. KS works with amplitude data, so that SqueeSAR uses amplitude as proxy for phase stability, an assumption that is also used in the amplitude dispersion index and other PSC selection methods.

Each DS is characterized statistically via the covariance matrix. By normalizing the amplitudes in the matrix to 1, a coherence matrix is obtained. The off-diagonal elements of the coherence matrix are estimates of the coherence values. For a true PS, the coherence matrix is a redundant singular matrix. This is not the case for a DS. The filtering of the interferogram causes the spatially filtered phase values to be inconsistent and thus 3D phase unwrapping becomes impossible. In SqueeSAR, the optimal phase values are derived using the coherence matrix via the phase triangulation algorithm (PTA) (Guarnieri & Tebaldini, 2008). Using PTA, N optimal phase values are obtained, where N is the number of slave images in the stack.

PTA provides a link between PS and DS, making it possible to characterize a DS through N phase values rather than $N(N - 1)/2$. This allows pre-processed data to be used in the standard PSInSAR processing chain. This means that well tested standard processes can be deployed with just the SqueeSAR pre-processing steps added. Figure 8.3 shows the full-graph interferometric combination used by SqueeSAR. Figure 8.4 shows a example of SqueeSAR application.

The SqueeSAR algorithm consists of six steps:

1. Use DespecKS to identify the SHP of each pixel.
2. Define DS all those pixels for which the number of SHP is larger than a given threshold.
3. Estimate the sample coherence matrix for all DS.
4. Apply the PTA algorithm to each coherence matrix.
5. Select the DS exhibiting PTA value higher than a certain threshold and use the N optimal phase values.

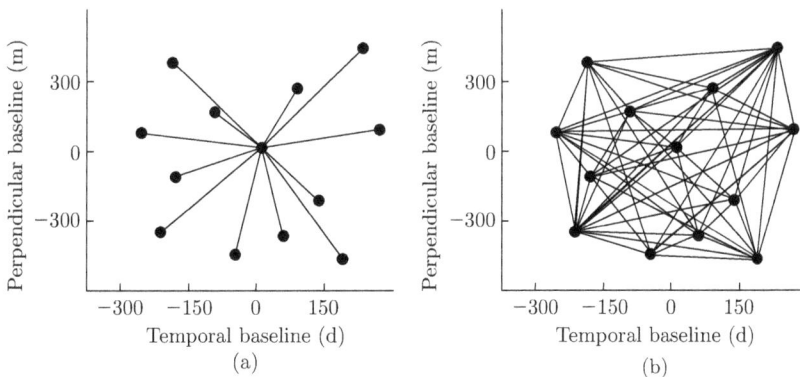

Figure 8.3 Combination of interferograms, with the temporal baseline in days on the x-axis and the perpendicular baseline on the y-axis. (a) star graph; (b) full graph used in SqueeSAR.

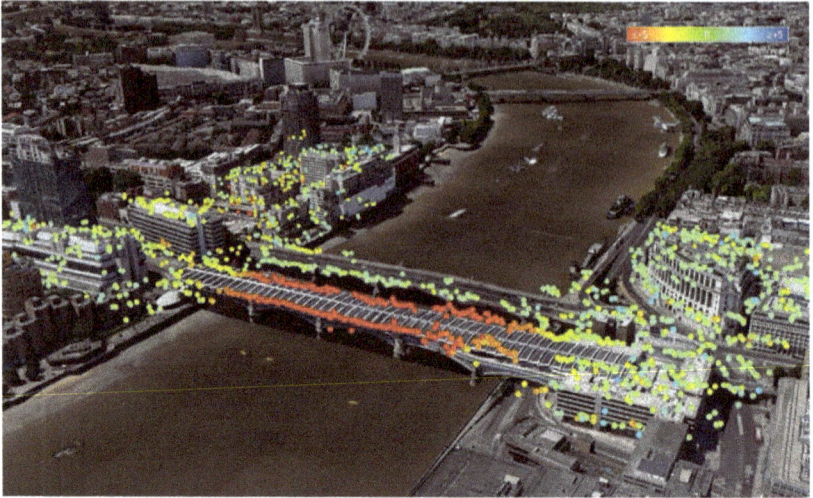

Figure 8.4 Infrastructure stability analysis with SqueeSAR (Courtesy of TRE ALTAMIRA).

6. Process the selected DS and PS together using the traditional PSInSAR algorithm.

SqueeSAR has therefore several advantages. By using an adaptive filtering, DespecKS, statistically homogenous pixels are combined into a DS, while the PS remain undisturbed. The optimal phase values from all interferograms are estimated using PTA, which then derives N optimal phase values. The term SqueeSAR is derived from the 'squeezing' of the N optimal phase values out of all possible interferograms. These values can then be used in a way similar to PSInSAR, allowing for 3D phase unwrapping and the use of the well-tested standard PSInSAR processing chain.

Chapter 9

Pixel Tracking and Point-Target Offset Tracking

Similar to DEM generation with SAR, there are phase-based and amplitude-based SAR approaches for surface motion estimation. The amplitude-based approaches are more stable and less affected by noise and temporal decorrelation, but are also less precise; the achievable precision directly correlates with the spatial resolution of the system.

Amplitude-based approaches are not bound to the wrapping problems as are phase-based methods and can therefore support much higher velocities. They are also not limited to measurements only in range direction. As a rule of thumb, we can therefore say that amplitude-based approaches are suitable for faster velocities as well as for motions in azimuth direction, whereas phase-based methods can achieve a much higher precision for slow-moving surfaces.

Similar to PS- and DS-InSAR in interferometry, we can also separate methods for point-like scatterers and distributed scatterers (DS) in amplitude-based surface motion estimation. There are several names for these approaches. We will herein refer to pixel tracking for the general methods and point-target offset tracking (PTOT) for methods focusing only on point scatterers.

The very first application for pixel tracking was published in 1991 for sea ice monitoring using Seasat SAR images (McConnell *et al.*, 1991). Scambos *et al.* (1992) used it to measure the velocities of glaciers. Pixel tracking is very often used for applications in sea ice and glacier monitoring, as both move too fast for InSAR and ice shows very low coherence, further hindering phase-based approaches. Pixel tracking can handle the fast motion of glaciers, especially sea ice.

Pixel tracking is not limited to these applications though. It can also be used, for example, in the estimation of earthquake-induced motion, as Michel *et al.* (1999) demonstrated, also using ERS data from the Landers earthquake, as did Massonnet *et al.* (1993) in their ground-breaking publication on differential SAR interferometry. The large displacement in the 2005 earthquake in Kashmir was also reconstructed using pixel tracking with ASAR data (Pathier *et al.* 2006).

The availability of high-resolution SAR data increases the applicability of amplitude-based methods. As the precision of pixel tracking and other amplitude-based methods depends on the spatial resolution, high-resolution data are of utmost importance. With the availability of high-resolution data, pixel tracking becomes more attractive for various applications, for example, in geology (Ruch *et al.*, 2016) or landslide detection (Milillo *et al.*, 2017; Singleton *et al.*, 2014). The technique also allows the monitoring of surface deformation caused by nuclear tests.

9.1   Pixel Tracking

In pixel tracking, motion is estimated from the differences in SAR amplitude images acquired at different times. In many ways, pixel tracking is to stereo radargrammetry as D-InSAR is to InSAR.

As in stereo-radargrammetry, the first step is to find homologous points in both images, which are pixels in both images that represent the same object on the ground. In stereo radargrammetry, we assume differences in the position between the two images are caused by differences in the acquisition geometry and therefore by height differences. In pixel tracking, we assume that these differences are instead caused by motion of the object on the ground between the acquisitions.

In reality, both can occur at the same time as the visible shift of homologous points can be caused by both: differences in acquisition geometery and object motion on the ground. So, similar to SAR interferometry, it is essential to separate both effects. The key is, similar to interferometry, the spatial baseline between the acquisitions. Stereo-radargrammetry, requires a large angular difference between the acquisition geometries to precisely estimate the height. While measuring motion differences instead of height differences, the angular differences between the acquisitions should therefore be minimized.

For pixel-tracking, images acquired at different times from the same orbit are therefore used, whereas for stereo-radargrammetry, images from different orbits are used.

The advantages and disadvantages of pixel tracking as compared to differential InSAR are similar to the differences between InSAR and stereo radargrammetry. Pixel tracking is less affected by atmospheric effects and temporal decorrelation, but the achievable accuracy depends strongly on the spatial resolution. Furthermore, pixel tracking is not limited by phase wrapping and therefore able to measure motion unambiguously even for fast motions. Another advantage of pixel tracking is the ability to measure motion in the azimuth direction as well.

Pixel tracking is therefore advantageous for measuring fast motions and motions in areas suffering from temporal decorrelation or motion that is predominantly in the azimuth direction. Pixel tracking, however, requires high-resolution data for measurements.

Pixel tracking is a relative measurement, measuring the motion differences between pixels rather than to a reference point that is assumed stable.

The first step in pixel tracking is therefore some form of co-registration between the images. The standard coarse-to-fine approach can be used. However, moving areas should be avoided in the co-registration. Orbit-based co-registration is advantageous, as it avoids using moving targets in the co-registration. Nevertheless, when using coarse-to-fine co-registration and assuming an affine transformation between the images, the influence of the moving pixels on the overall transformation will be minimized.

The co-registration should then be validated using a reference point. Since the reference point is assumed to be stable, the co-registration of the reference point should be at sub-pixel precision. To validate this within the highest possible precision, the reference point is ideally a point target with a high SCR. It is also preferable to use several point targets that are not moving over the image to validate the co-registration of the image in the non-moving image parts.

It is also possible to avoid the co-registration and resampling of the slave image, by just identifying the reference point in both images, ideally to sub-pixel precision. This requires both images to be geometrically stable and identical except for the shift in range and azimuth between the images. This assumption can be made with most SAR images due to the high geometrical precision of SAR. However, small differences in pixel sizes are not uncommon, especially in spotlight image acquisitions, so that it is again preferable to validate the methods by including several known stable point targets and establish their stability to avoid errors in the processing.

Afterwards, the motion is estimated for each pixel using a measurement of similarity on oversampled image chips, as shown in Figure 9.1. Normalized cross-correlation is a common approach, but other similarity measurements are also usable. For example, by including the phase, the coherence can be used. This so-called coherence-tracking allows for higher precision, but requires phase stability and is therefore not suitable in all applications.

As in stereo radargrammetry, a measurement of the overall fitting is provided, allowing us to filter for estimated motions that are reliable enough. The search area, the size of the mask used for comparison, as well as the oversampling factor need to be defined before pixel tracking. Larger search area increases the calculation time and also allows for the detection of larger (faster) motion. Larger oversampling also increases the calculation time, but can increase the precision up to a limit. Larger masks also increase calculation time, while also increasing the precision. Masks too large though can include pixels with very different motion parameters, which would disturb the precise estimation. Therefore, the mask size should not be larger than the moving objects. With landslides as an example, the mask size should be smaller than the size of the landslide.

In terms of the achievable precision, there is no theoretical estimation for the general pixel-offset tracking method available. For

(a)

(b)

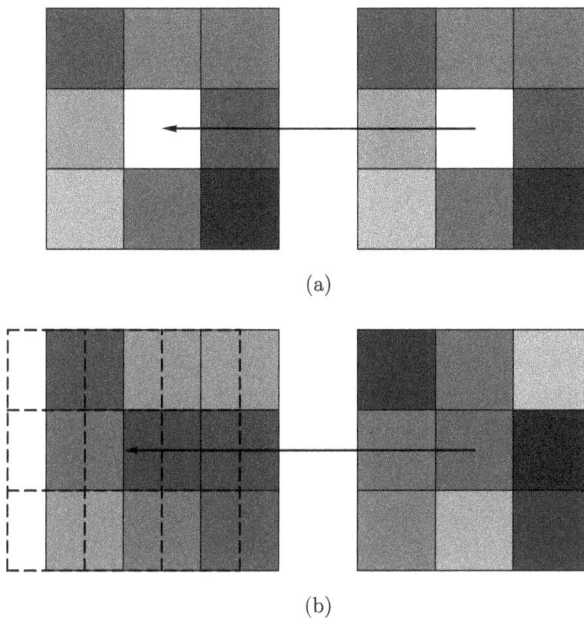

Figure 9.1 Pixel tracking of a stable point (a) and a pixel shifting half a pixel to the left (b).

coherence-tracking though, the achievable accuracy is given by (Bamler & Eineder, 2005):

$$\sigma_{CR} = \sqrt{\frac{3}{2N}} \frac{\sqrt{1 - \gamma^2}}{\pi\gamma} \qquad (9.1)$$

where σ_{CR} is the standard deviation of the pixel offset error, N is the number of samples in the estimation window and γ is the coherence. For standard cross-correlation-based pixel tracking, the error is worse and at least $\sqrt{2}$ larger. As shown by Bamler and Eineder (2005), the error is expected to be even bigger for lower coherences. So, even for high correlation values, the sub-pixel accuracy is limited, and in general, we can assume a 0.1–1 pixel accuracy.

As the error is expressed in pixel, the direct relation between the spatial resolution and the error becomes apparent. Assuming a standard deviation of the error of half a pixel would relate to about 0.4 m for a high-resolution spotlight image, but up to 10 m for a mid-resolution SAR system, like Sentinel-1 in azimuth direction.

9.2   Point-Target Offset Tracking

Point-target offset tracking (PTOT) is another approach for pixel tracking, focusing on point targets. It is therefore similar to PSI in comparison to D-InSAR. For point targets where, similar to permanent scatterers (PS), only one strong scatterer is dominating the backscattering of a resolution cell, the sub-pixel position of that scatterer can be estimated at a much higher precision. The accuracy can be estimated by (Bamler & Eineder, 2005)

$$\sigma_{\text{point}} = \frac{\sqrt{3}}{\pi} \frac{1}{\sqrt{\text{SCR}}} \qquad (9.2)$$

with SCR being the signal-to-clutter ratio of the target. The precision depends then solely on the SCR. There is no estimation window to be considered, as multi-looking or averaging is not advised to keep the original point-target measurement.

The precision as a function of the SCR is shown in Figure 9.2. For targets with an SCR of 20 dB, the accuracy is about 1/20th of a pixel. Point targets with high SCR can therefore be positioned with a high sub-pixel accuracy, which, for example, is used in image co-registration (Serafino, 2006). The ability for such precise sub-pixel positioning is also treated in the following chapter on SAR geodesy.

The precise sub-pixel position is determined from an oversampled image chip by estimating the position assuming the sinc function as point response function of the point scatterer. Following the position of point scatterers, differences in the positioning of point scatterers between images can also be used to derive motion (Hu *et al.*, 2013).

For high-resolution spotlight data, relative motions with about 5 cm or better can be followed for targets with about 20 dB SCR. Assuming the

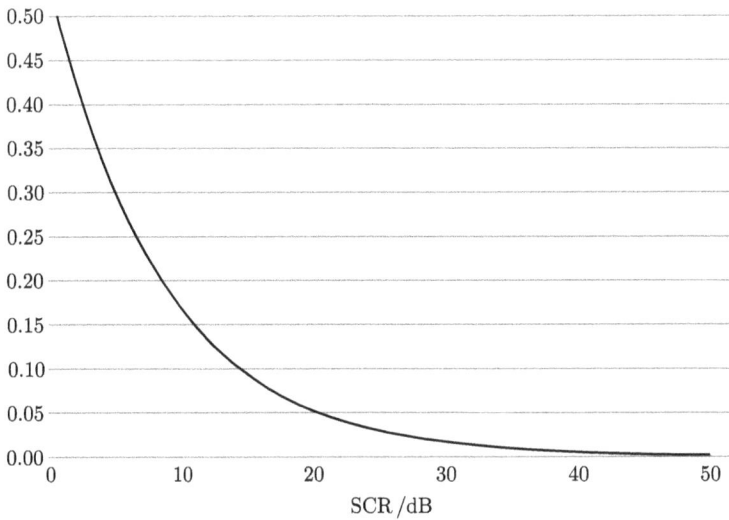

Figure 9.2   Sub-pixel position precision as function of the SCR.

same 1/20th of a pixel accuracy, about 25 cm motion in range and 1 m motion in azimuth could be followed using Sentinel-1 data. Significantly higher accuracies can be achieved using higher SCR.

Using high-resolution spotlight data, the accuracy can reach the centimeter level, allowing measurements for many applications. For faster motions, like, for example, landslides, PTOT can be very useful. However, PTOT requires high SCR targets. Similar to the problems of PS density, such targets are typically not often found outside of urban areas. Therefore, the applicability is often limited by the lack of point targets in the area of interest.

Using artificial targets on areas of interest can overcome this problem, but requires the installation of such targets. This can lead to myriads of new problems, including the theft or damage of such targets or the possible change in an active landslide deformation pattern by the installation of heavy corner reflectors.

Chapter 10

Motion from SAR Geodesy

The term "SAR geodesy" is sometimes used to describe different approaches for precise measurements with SAR. In this book, we use the term to describe the precise measurement of absolute 3D world coordinates from SAR images. The keyword here is *absolute*. The approaches described in the previous chapters allowed for the precise measurement of relative height differences or relative motion. SAR geodesy offers precise absolute positioning. SAR geodesy requires a well-calibrated SAR system and precise orbit parameters.

The approach was first described by Eineder *et al.* (2011) and then extended by Cong *et al.* (2012). These first approaches demonstrated the high-precision georeferencing ability of TerraSAR-X. This was further extended to the absolute measurement of 3D coordinates (Gisinger *et al.*, 2015) using stereo configurations and applied toward the automated detection of ground-control points in automated mapping (Montazeri *et al.*, 2018). Absolute precision in the centimeter to decimeter range can be achieved with this approach.

10.1   Positioning in SAR Images

The axis in an SAR image relates to the timing in azimuth and range of the image, also called slow time and fast time. The coordinates in the range direction of the SAR image gives the travel time of the signal. The pixel in the first column in range has a defined travel time, which is increased for every pixel. In azimuth direction, the pixel also represents

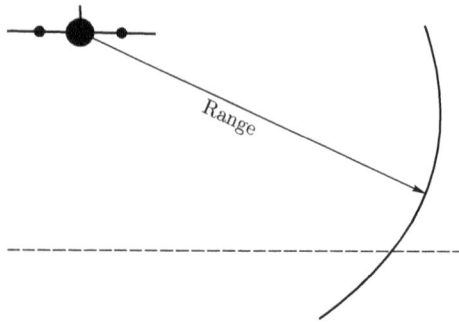

Figure 10.1   Estimating real-world positions from SAR coordinates.

a time, the time difference of the so-called zero-Doppler time from a given reference time. From the x/y image coordinate of an SAR image, the travel time of the signal in range and the timing of the acquisition in azimuth can be calculated. From the timing information in azimuth, the exact position of the sensor for a given azimuth line can be derived using the precise orbit information. Therefore, we can derive the sensor position as well as the distance of the point from the sensor from the image coordinates.

This principle is shown in Figure 10.1. Knowing the sensor position, the direction of travel and the looking direction of the sensor (left or right), the position can be reduced to a quarter circle of a given radius (range) from a given position. However, the position on the ground remains ambiguous, but is somewhere on that circle. Knowing the precise height, e.g. from a DEM, the position can be retrieved in world coordinates.

Positioning is therefore an ambiguous process. Having a DEM, or assuming a constant height for all pixels, georeferencing in SAR images is typically based on an iterative process (Curlander, 1982).

10.2   Factors Influencing the Positioning

The precision of the aforementioned process depends strongly on knowledge of the correct height. Furthermore, it requires precise orbit information, so that the timing information from the SAR image is translated into a correct position of the sensor. Typically, the speed of light in vacuum is assumed for the calculation of the distance from the range time. As the signal travels through the atmosphere, this assumption is clearly not correct and needs to be adjusted. Most of these factors are well known in geodesy, but are often not taken into consideration for SAR.

10.2.1   Static Influencing Factors

A certain understanding of the meaning of the different coordinate systems is essential. Using a geographical coordinate system, the coordinates are given in latitudes and longitudes, while the height is given in meters. Typically, the height is given with respect to the ellipsoid that is used to best represent the Earth. Often, WGS84 is used, which is the ellipsoid of the World Geodetic System from 1984. The Earth is not exactly an ellipsoid, but rather a Geoid and the real height of the Geoid can vary significantly from the ellipsoidal height. Therefore, the given height might need to be corrected to Geoid height if given as ellipsoidal height. The height of the Geoid for each point of the Earth can be derived from several sources, with freely available information.

Another important factor is the orbit information. For precise positioning, the most precise orbit information available should be used. Additionally, azimuth timing errors should be corrected. This can be part of the calibration of the SAR system and might be included in the metadata. For example, the DLR provides azimuth timing error information as additional information in the metadata of TerraSAR-X.

As discussed, an electromagnetic wave does not travel with the speed of light in vacuum through the Earth's atmosphere. The reduction of the speed can be divided into two parts. There is a constant factor slowing down the signal, also called the path delay, coming from the dry atmosphere. Additionally, there is a dynamic factor caused by the changing water-vapor content and the air pressure in the atmosphere, which also influence the speed of the electromagnetic wave. These factors will be discussed in more detail in the following section.

The constant factors can cause errors in the meter or several meters range and can be corrected relatively easily. The dynamic factors described in the following section are in the decimeter range and can be more difficult to correct.

10.2.2   Dynamic Influencing Factors

Although we conceive world coordinates as something static, they also have a dynamic element. On our dynamic Earth, nothing is really stable and our coordinates are moving with the plate tectonics, several centimeters per year. Therefore, the coordinates are given relative to a reference point or reference frame. To account for the relative shift over time, precise coordinates are assigned to a reference frame and a time, the so-called epoch. In this way, the precise absolute location can be determined.

In terms of SAR geodesy, the reference frame and epoch should be known, and transformed into the reference frame and epoch of the coordinates used in describing the sensor positions when gathering point coordinates. The differences can be significant. A typical reference frame is the European Terrestrial Reference Frame: ETRF89. With the reference now more than 30 years old and assuming a motion of 2 cm · yr^{-1}, differences of 0.5 m can be expected, far too much variation when absolute positioning in the centimeter to decimeter range is desired. Fixing this is relatively simple though, as the coordinates just need to be transformed. The key to the problem is awareness, as many are not aware of the issue.

Another factor that must be considered in absolute positioning is the solid Earth Tides. We know the tides from the sea. The positions of the Moon and the Sun (and other gravitational bodies) affect land masses as well, leading to changes in their absolute position in the horizontal and vertical directions, as shown in Figure 10.2. These changes are in the decimeter range and therefore need to be corrected. The information on

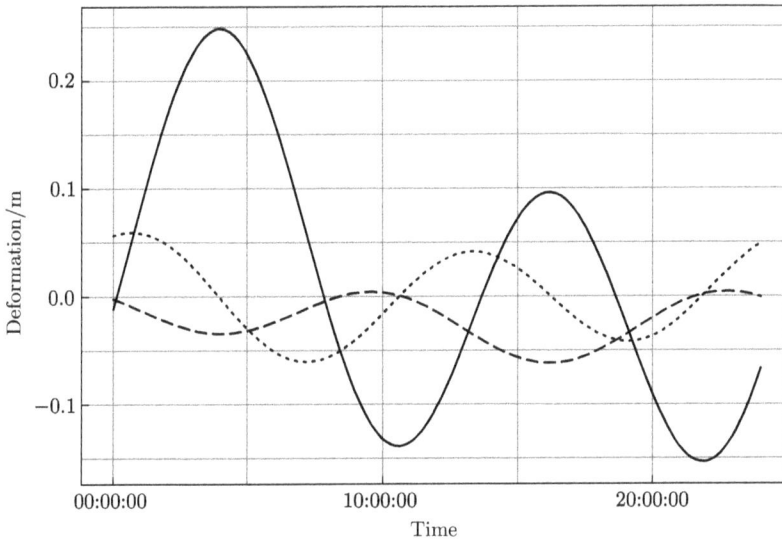

Figure 10.2 Solid Earth tides in Wuhan, China, on 2020-09-01. Earth tides to north are shown as dashed line, Earth tides to east are shown as dotted line and the vertical Earth tides are shown as solid line.

the Earth tides can be received for a given time and position, and can then be corrected accordingly.

Another contributing factor is the atmosphere, especially the troposphere and ionosphere. We discussed the atmosphere already under the static influencing factors. We can separate the atmospheric path delay into a static or constant part and a dynamic part. The dynamic part depends on several factors, with the water-vapor content, the temperature, and the air pressure being the dominant dynamic factors. In addition to the troposphere, the total electron content (TEC) in the ionosphere also delays the signal. This part of the signal delay depends then on the wavelength, affecting long-wavelength systems stronger than shorter-wavelength systems. The overall path length extension L can be calculated as

$$L = 10^{-6} \int_{r_{\text{ground}}}^{r_{\text{sensor}}} k_1 \frac{P_{\text{d}}}{T} + k_2 \frac{e}{T} + k_3 \frac{e}{T^2} + 1.45 W_{\text{cl}} + 4.028 \frac{n^2}{f^2} dr \quad (10.1)$$

where k_1, k_2, and k_3 are constants, T is the temperature, e is the partial pressure of the water vapor, W_{cl} is the liquid water, n is the electron density and f is the radar frequency. These values can be derived from other remote sensing sources, like MERIS, or from numerical weather models. As the path delay also affects GNSS, these are well studied and methods from GNSS and navigation can be applied.

10.2.3   Sub-Pixel Positioning Precision

To achieve high positioning accuracies in the centimeter range, the backscattering object should be a dominant point scatterer similar to the previously discussed PS points or the targets in the point-target offset tracking. Having such a dominant point scatterer, the theoretical sub-pixel precision is (Bamler & Eineder, 2005)

$$\sigma_{point} = \frac{\sqrt{3}}{\pi} \frac{1}{\sqrt{SCR}} \tag{10.2}$$

with SCR being the signal-to-clutter ratio of the target. For targets with a signal-to-clutter ratio of 20 dB, the accuracy is about $1/20$th of a pixel (see also Figure 9.2 of Chapter 9).

The SCR depends on the background noise and the size of the target. In Figure 10.3, the relation between the achievable sub-pixel positioning and the size of a trihedral corner reflector is shown. For smaller wavelengths, the corner size can be smaller. The spatial resolution of the sensor is also important, as smaller resolution cells include less clutter and therefore can significantly improve the SCR.

The overall positioning precision therefore depends strongly on the spatial resolution of the system. High spatial resolution increases the SCR. As the precision is then expressed as a fraction of the pixel size, the achievable measurement precision then directly relates to the spatial resolution of the system again. Other factors include the orbit accuracy of the system, and the ability to precisely derive the geodynamical parameters for the correction of the factors influencing the positioning precision.

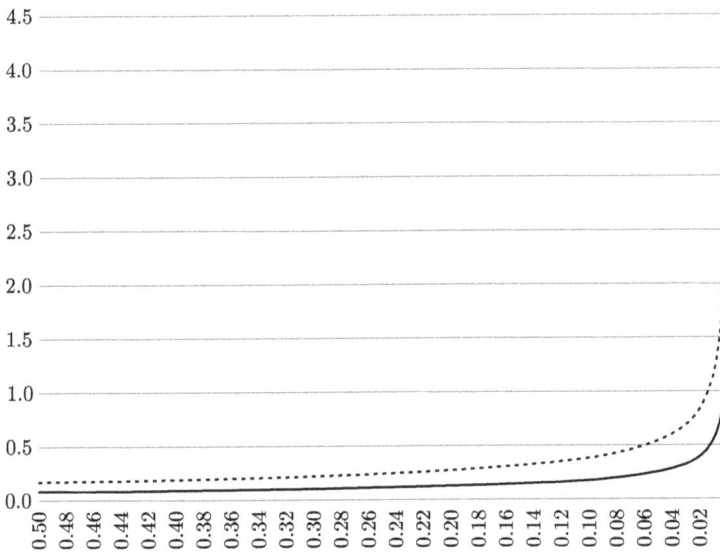

Figure 10.3 Sub-pixel precision as function of the corner size with an assumed background noise of −1 dB. Solid line denotes TerraSAR-X stripmap data and dotted line denotes Sentinel-1 data.

10.3 Absolute 3D Positioning Using Stereo Configuration

From the previous descriptions, a point scatterer of a known position and height can be precisely positioned in an SAR image. However, this is of only limited practical use, as the precise height of the target is needed. The application value lies in deriving the precise 3D position from SAR image coordinates. However, the 3D position is ambiguous, as shown in Figure 10.1. To derive a 3D position from a 2D coordinate system, additional information is necessary, for example, the height. Alternatively, a second SAR image acquired from a different position can be used. As long as an identical object can be discerned in both images, the precise location can be derived. As shown in Figure 10.4, the position can be identified at the intersection of the range circles.

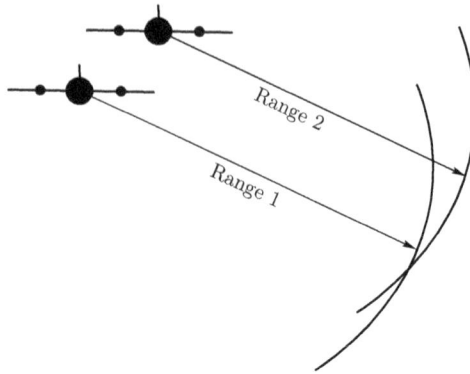

Figure 10.4   3D positioning using SAR stereo configuration.

Figure 10.4 simplifies the issue though, as instead of two circles intersecting in 2D, we have a three-dimensional intersection of spherical elements that may have ambiguous solutions. For a better solution, it may therefore be beneficial to include more than two images in the estimation.

10.4   Absolute Methods versus Relative Methods

At this point, it might be useful to elaborate on the differences of SAR geodesy as a method of absolute measurement in comparison to all previous methods that provide relative measurement. After seeing the different static and dynamic error sources, one may wonder why, for example, Solid Earth Tides have not been discussed before. Decimeter-level changes would significantly affect the precise interferometric measurements dealing with path changes in the millimeter range. Dry atmosphere causes errors in the meter range, but why has this not been mentioned before?

In relative measurements, we take the range (or height) differences between pixels. That is to say, we measure, for example, the phase difference between two pixels and estimate the height and motion differences between these pixels. In such a case, any important factor that influences both pixels in a similar way is irrelevant. As the solid Earth Tides are approximately similar for all pixels in an image, there is no phase difference between two pixels induced from that. Similarly, the atmospheric phase screen in InSAR is the difference in the atmospheric path delay between two pixels. As the dry atmospheric path delay is similar all over the image, it is irrelevant for InSAR, but the wet atmosphere has a significant effect, as it changes in time and in space.

Relative methods though need a reference such as one at a known height for DEM generation or one assumed to be stable for motion estimation. As we saw from the different error sources in this chapter, there is no stability on our dynamic Earth. For most applications, that does not matter, as we are mostly interested in relative motion differences. The motion of the solid Earth Tides do not affect us much, nor does uplift and sinking on a continental scale. Plate tectonics strongly influence us in general, but only local differences in motion occur, along fault lines and sometimes expressed in earthquakes matter. We are interested in relative motion differences, so that the requirement for a presumably stable reference point is not a huge drawback.

Absolute measurements do not need a reference point, although each coordinate system again refers to some reference frame. Free measurement is a powerful and unique capability of SAR geodesy, demonstrating the inherent precision of SAR sensors as geodetic measurement tools.

10.5   Motion from SAR Geodesy

With the ability to measure positions in the centimeter to decimeter range, motions exceeding the error interval can be measured. As a series of measurements over time, linear motion can be estimated at higher precision, supposedly in the centimeter range. The relative measurements used in point-target offset tracking have less variables to consider though and are therefore less error-prone. As long as relative measurements of the motions are suitable, that is, as long as we can define a relatively stable reference point, these measurements are more suitable for motion differences. SAR geodesy can then assist in calculating absolute values, for example, by measuring the motion of the supposedly stable reference point.

However, for some motions, we cannot define a stable reference point. These are typically motions affecting a wider area. Examples would be the isostatic rebound of land masses. Motions of islands, being tectonic motions, subsidence, or uplift, are also examples of possible applications, as there might be no suitable reference close enough to realize a relative motion measurement.

For such applications, motion from SAR geodesy can be an alternative. The achievable precision of a few centimeters can be sufficient if longer time series of SAR geodetic measurements are used, which can in turn also increase the overall precision with averaging for easier outlier detection.

10.6   SAR Geodesy in Practice

In practice, using SAR geodesy requires careful processing and quality assessment as a high precision is required, but the available input data for error correction is sometimes less precise and more erroneous than desired. This is especially true for the weather models, which can increase the precision, but can also contain large outliers and, in such cases, could even decrease positioning accuracy.

10.6.1   Finding Suitable Points

The first problem in SAR geodesy is finding suitable point-scatterers. In many SAR geodesy experiments, corner reflectors are used. In such cases, point scatterers with a high SCR are available at already known positions. However, for most applications, we are interested in finding suitable points without setting up artificial targets.

Points suitable for SAR geodesy need to have a high SCR to allow for a precise sub-pixel positioning of the scattering center. Furthermore, the scatterer needs to have a high SCR under different incidence angles and possible different orbit directions. Scatterers that keep strongly scattering from different angles are far more seldom than strong scatterers that are stable from one looking direction. Compared to the search for PS points in PSI processing, it is therefore more difficult to find a high number of suitable candidates.

The situation is shown in Figure 10.5. The corner reflector, dihedral or trihedral, reflects strongly under different incidence angles. However, a corner will not reflect looking from the opposite direction as in cross-heading orbits, which is looking from ascending and descending orbits. On the other hand, a pole will reflect from different incidence angles and from different looking directions. The two corner reflectors used in this experiment are shown in Figure 10.6.

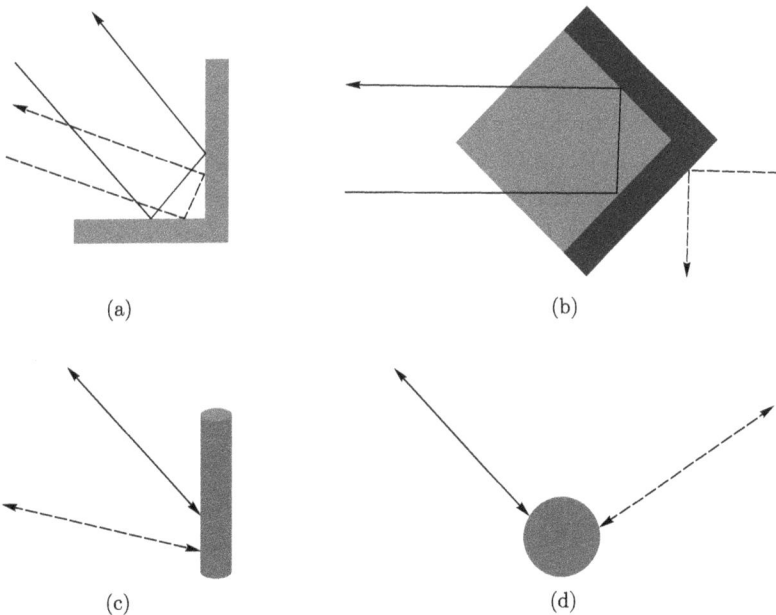

(a) (b)

(c) (d)

Figure 10.5 Reflections from artificial targets: (a) corner reflector reflection from co-heading orbits; (b) corner reflector reflection from cross-heading orbits; (c) pole reflection from co-heading orbits; (d) pole reflection from cross-heading orbits.

Figure 10.6   The two corner reflectors used at the SAR geodesy experiment.

In terms of accuracy, we prefer cross-heading orbits and ideally multiple cross-heading orbits to have the best stereo effect for the measurement. In terms of available points, the situation is very different. Analyzing four TerraSAR-X high-resolution spotlight images at a spatial resolution of about 0.8 m, we can find 8741 points with an SCR of 15 dB or higher from two ascending orbits. This relative high point density, as shown in Figure 10.7, allows for an acceptable density of measurements. Combining ascending and descending orbits though drastically reduces the point density to only 168 points that have a high SCR in all four images, with two images from an ascending orbit and two images from a descending orbit.

Finding enough high SCR points is a big problem for SAR geodesy, especially points which allow cross-orbit processing. This is almost limited exclusively to poles and they often do not provide a very high SCR.

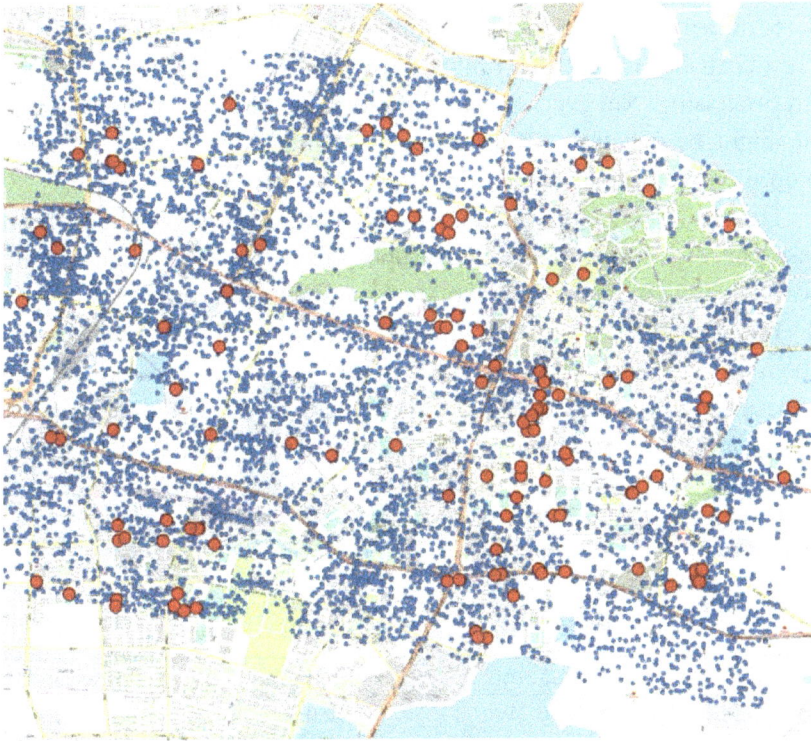

Figure 10.7 8741 high SCR points found in two ascending orbits (blue) and 168 points in two ascending and two descending orbits (red) using TerraSAR-X high-resolution spotlight data over Wuhan.

10.6.2 Correction for Geodynamical Effects

In our example, we separate the general geodynamical effects from the dynamical (or wet) atmosphere effects. Including the basic geodynamical effects in the processing is rather simple and immediately improves the precision. The very first task in processing, as it may have the biggest effect, is to make sure to correct the processed height with Geoid information to fit into the reference frame of the satellite.

The next effect to be corrected for is the Solid Earth Tides. The extent of the Solid Earth Tides can be calculated for the acquisition time and the position and then the positions can be adjusted accordingly.

SAR azimuth timing error correction would be the next step. TerraSAR-X provides SAR timing error information, which can be used in processing. Not every sensor provides this information however, so it might be necessary to estimate the timing error based on known coordinates in the image to be processed.

Furthermore, a constant value for the path delay in the atmosphere, especially the ionosphere and troposphere, can be estimated and corrected. Again, TerraSAR-X offers a constant correction value.

Based on this information, points can be measured at a comparably high absolute precision already. In an experiment based on the corrections described here, and without using a weather model, absolute positioning better than 0.5 m can be achieved (Wang *et al.*, 2016). Tables 10.1 and 10.2 show the results of the experiment we conducted in Wuhan, China, based on this dry-atmosphere approach.

Table 10.1   Using two TerraSAR-X high-resolution images from an ascending orbit.

Point	Δ East	Δ North	Δ Height
B405	−0.25 m	−0.033 m	0.31 m
P066	−0.30 m	−0.062 m	0.343 m

Table 10.2   Using two TerraSAR-X high-resolution images from a descending orbit.

Point	Δ East	Δ North	Δ Height
B405	0.1 m	−0.005 m	0.231 m
P066	0.05 m	0.028 m	0.333 m

From these results, we can see that the results in the Northing direction at a sub-decimeter precision. This is because the error in Northing is dominated by errors in azimuth, related to azimuth timing errors and can be corrected. The errors in Easting and height are dominated by range errors, linked to path delay errors and the wet atmosphere, which were not corrected here.

10.6.3 Correction for Dynamic Atmospheric Effects

The correction of the atmosphere requires the values for the temperature T, the partial pressure of the water vapor e, and the liquid water W_{cl}. With this, we can calculate the path delay

$$L_{tropo} = 10^{-6} \int_{r_{ground}}^{r_{sensor}} k_1 \frac{P_d}{T} + k_2 \frac{e}{T} + k_3 \frac{e}{T^2} + 1.45 W_{cl} \qquad (10.3)$$

In our example, we ignored the influence from the ionosphere as it is rather small in the X-band. Thus, we corrected the ionosphere based only on the constant given in the TerraSAR-X data.

The values for T, e, and W_{cl} can be derived from available weather models. There are several models available. In this example, we used the ERA Interim model and the Merra-2 model. All of these models have a coarse spatial resolution, with ERA Interim having a spatial resolution of about 80 km and Merra-2 having about 50 km. This resolution is very low compared to the high-resolution SAR images.

The main issue in using these weather models for SAR geodesy is dealing with the low resolution. There are also weather models with higher resolutions available, but their data are not freely available. Nevertheless, for better results, these models should be considered.

An accurate interpolation is necessary, to mitigate the problems from the coarse resolution of the data. In this regard, the following experiments do not show the full possibilities of SAR geodesy, as better results can be achieved. Nevertheless, they provide us with insight into the problems that can arise from weather models.

For the tests, we estimated the path delay based on the ERA Interim and Merra-2 data for an ascending and a descending orbit, each with two TerraSAR-X high-resolution spotlight images over Wuhan (Tables 10.3–10.6).

We can see huge differences between the ERA Interim and the Merra-2 results. We also see differences to the dry atmosphere approach. Here, we see an improvement, especially in the height estimation. In Easting, we see very good results in the descending data in ERA Interim and Merra-2, and reach very high accuracy in the ERA Interim case.

Table 10.3   Using two TerraSAR-X high-resolution images from an ascending orbit corrected using ERA Interim.

Point	Δ East	Δ North	Δ Height
B405	−0.66 m	−0.11 m	−0.18 m
P066	−0.37 m	−0.09 m	0.08 m

Table 10.4   Using two TerraSAR-X high-resolution images from a descending orbit corrected using ERA Interim.

Point	Δ East	Δ North	Δ Height
B405	0.03 m	0.03 m	−0.04 m
P066	−0.02 m	0.01 m	0.06 m

Table 10.5   Using two TerraSAR-X high-resolution images from an ascending orbit corrected using Merra-2.

Point	Δ East	Δ North	Δ Height
B405	−0.32 m	−0.07 m	0.05 m
P066	−0.17 m	−0.06 m	0.22 m

Table 10.6   Using two TerraSAR-X high-resolution images from a descending orbit corrected using Merra-2.

Point	Δ East	Δ North	Δ Height
B405	−0.09 m	0.05 m	0.03 m
P066	−0.13 m	0.04 m	0.11 m

In the ascending case, the situation is different and the overall accuracy in Easting is even below the dry atmosphere example. This is due to a high volatility of these models. They provide good results, but they can also have large outliers, especially in turbulent weather conditions; keeping in mind that the rather low resolution of the data cannot correctly model turbulent weather situations locally. Under such circumstances, the outliers of such measurements can be large.

Additionally, the presented example was processed with a non-optimal solution. The delay was estimated on the zenith path delay and then extrapolated to slant range. This works well in calm weather scenarios with only limited spatial differences in the path delay. A better way is to estimate the path delay of the atmosphere along the slant-range path. Using more than two images would not only allow for a better averaging of the sub-pixel positioning error but also can increase the possibility of detecting outliers caused by atmospheric models.

Bibliography

Adam, N., Kampes, B., & Eineder, M. (2004). The development of a scientific persistent scatterer system: Modifications for mixed ERS/ENVISAT time-series. ENIVSAT and ERS Symposium, 1–9.

Ansari, H., De Zan, F., & Bamler, R. (2017). Sequential estimator: Toward efficient InSAR time series analysis. *IEEE Transactions on Geoscience and Remote Sensing*, *55*(10), 5637–5652.

Ansari, H., De Zan, F., & Parizzi, A. (2020). Study of systematic bias in measuring surface deformation with SAR interferometry. *IEEE Transactions on Geoscience and Remote Sensing*, doi: https://doi.org/10.1109/tgrs.2020.3003421.

Balz, T., Zhang, L., & Liao, M. (2013). Direct stereo radargrammetric processing using massively parallel processing. *ISPRS Journal of Photogrammetry and Remote Sensing*, *79*, 137–146.

Bamler, R., & Eineder, M. (2005). Accuracy of differential shift estimation by correlation and split-bandwidth interferometry for wideband and delta-k SAR systems. *IEEE Geoscience and Remote Sensing Letters*, *2*(2), 151–155.

Berardino, P., Fornaro, G., Lanari, R., & Sansosti, E. (2002). A new algorithm for surface deformation monitoring based on small baseline differential SAR interferograms. *IEEE Transactions on Geoscience and Remote Sensing*, *40*(11), 2375–2383.

Blanco, P., Mallorqui, J. J., Duque, S., & Monells, D. (2008). The coherent pixels technique (CPT): An advanced D-InSAR technique for nonlinear deformation monitoring. *Pure and Applied Geophysics*, *165*(2008), 1167–1193.

Bruzzone, L., Marconcini, M., Wegmuller, U., & Wiesmann, A. (2004). An advanced system for the automatic classification of multitemporal SAR images. *IEEE Transactions on Geoscience and Remote Sensing*, *42*(6), 1321–1334.

Cafforio, C., Prati, C., & Rocca, F. (1991). SAR data focusing using seismic migration techniques. *IEEE Transactions on Aerospace and Electronic Systems*, *27*(2), 194–207.

Capaldo, P., Crespi, M., Fratarcangeli, F., Nascetti, A., & Pieralice, F. (2011). High-resolution SAR radargrammetry: A first application with COSMO-SkyMed Spotlight imagery. *IEEE Geoscience and Remote Sensing Letters*, *8*(6), 1100–1104.

Carrara, W. G., Goodman, R. S., & Majewski, R. M. (1995). *Spotlight Synthetic Aperture Radar: Signal Processing Algorithms (IPF)*. Boston, London: Artech House.

Chen, C. W., & Zebker, H. A. (2001). Two-dimensional phase unwrapping with use of statistical models for cost functions in nonlinear optimization. *Journal of the Optical Society of America A, 17*(3), 338–351.

Cong, X., Balss, U., Eineder, M., & Fritz, T. (2012). Imaging geodesy–centimeter-level ranging accuracy with TerraSAR-X: An update. *IEEE Geoscience and Remote Sensing Letters, 9*(5), 948–952.

Costantini, M. (1998). A novel phase unwrapping method based on network programming. *IEEE Transactions on Geoscience and Remote Sensing, 36*(3), 1–9.

Costantini, M., Falco, S., Malvarosa, F., Minati, F., Trillo, F., & Vecchioli, F. (2014). Persistent scatterer pair interferometry: Approach and application to COSMO-SkyMed SAR data. *IEEE Journal of Selected Topics in Applied Earth Observations and Remote Sensing, 7*(7), 2869–2879.

Crisp, D. J. (2004). *The State-of-the-Art in Ship Detection in Synthetic Aperture Radar Imagery*. Australian Government, Department of Defense.

Crosetto, M., Crippa, B., & Biescas, E. (2005). Early detection and in-depth analysis of deformation phenomena by radar interferometry. *Engineering Geology, 79*(1–2), 81–91.

Crosetto, M., Biescas, E., Duro, J., Closa, J., & Arnauld, A. (2008). Generation of Advanced ERS and Envisat Interferometric SAR Products Using the Stable Point Network Technique. *Photogrammetric Engineering and Remote Sensing, 4*(8), 443–450.

Crosetto, M., Monserrat, O., Cuevas-González, M., Devanthéry, N., & Crippa, B. (2016). Persistent scatterer interferometry: A review. *ISPRS Journal of Photogrammetry and Remote Sensing, 115*(100), 78–89.

Cumming, I. G., & Wong, F. H. (2005). *Digital Processing of Synthetic Aperture Radar: Algorithms and Implementations*. Norwood: Artech House.

Curlander, J. C. (1982). Location of spaceborne SAR imagery. *IEEE Transactions on Geoscience and Remote Sensing, GE-20*(3), 359–364.

Denos, M. (1992). A pyramidal scheme for stereo matching SIR-B imagery. *International Journal of Remote Sensing, 13*(2), 387–392.

Devanthéry, N., Crosetto, M., Monserrat, O., Cuevas-González, M., & Crippa, B. (2014). An approach to persistent scatterer interferometry. *Remote Sensing, 6*(7), 6662–6679.

Eineder, M., Adam, N., Bamler, R., Yague-Martinez, N., & Breit, H. (2009). Spaceborne spotlight SAR interferometry with TerraSAR-X. *IEEE Transactions on Geoscience and Remote Sensing, 47*(5), 1524–1535.

Eineder, M., Minet, C., Steigenberger, P., Cong, X., & Fritz, T. (2011). Imaging geodesy — Toward centimeter-level ranging accuracy with TerraSAR-X. *IEEE Transactions on Geoscience and Remote Sensing, 49*(2), 661–671.

Eldhuset, K. (2004). An automatic ship and ship wake detectio system for spaceborne SAR images in coastal regions. *IEEE Transactions on Geoscience and Remote Sensing, 34*(4), 1010–1019.

Farr, T. G., Rosen, P. A., Caro, E., Crippen, R. E., Duren, R. M., Hensley, S., Kobrick, M., Paller, M., Rodriguez, E., Roth, L. E., et al. (2007). The shuttle radar topography mission. *Reviews of Geophysics, 45*(2). RG2004.

Fayard, F., Meric, S., & Pottier, E. (2007). Matching stereoscopic SAR images for radargrammetric applications. *IEEE International Geoscience and Remote Sensing Symposium*, 4364–4367.

Ferretti, A., Fumagalli, A., Novali, F., Prati, C., Rocca, F., & Rucci, A. (2011). A new algorithm for processing interferometric data-stacks: SqueeSAR. *IEEE Transactions on Geoscience and Remote Sensing, 49*(9), 3460–3470.

Ferretti, A., Prati, C., & Rocca, F. (2000). Nonlinear subsidence rate estimation using permanent scatterers in differential SAR interferometry. *IEEE Transactions on Geoscience and Remote Sensing, 38*(5), 2202–2212.

Ferretti, A., Prati, C., & Rocca, F. (2001). Permanent Scatterers in SAR Interferometry. *IEEE Transactions on Geoscience and Remote Sensing, 39*(1), 8–20.

Forbes, N., & Mahon, B. (2014). *Faraday, Maxwell, and the Electromagnetic Field: How Two Men Revolutionized Physics*. Amherst: Prometheus Books.

Frost, V. S., Stiles, J. A., Shanmugan, K. S., & Holtzman, J. C. (1982). A model for radar images and its application to adaptive digital filtering of multiplicative noise. *IEEE Transactions on Pattern Analysis and Machine Intelligence, PAMI-4*(2), 157–166.

Gabriel, A., Goldstein, R. M., & Zebker, H. A. (1989). Mapping small elevation changes over large areas: Differential radar interferometry. *Journal of Geophysical Research, 94*, 1919–1983.

Gabriel, A. K., & Goldstein, R. M. (1988). Crossed orbit interferometry: Theory and experimental results from SIR-B. *International Journal of Remote Sensing, 9*(5), 857–872.

Ghiglia, D. C., & Romero, L. A. (1996). Minimum L^p-norm two-dimensional phase unwrapping. *Journal of the Optical Society of America A, 13*(10), 1999–2013.

Gisinger, C., Balss, U., Pail, R., Zhu, X. X., Montazeri, S., Gernhardt, S., & Eineder, M. (2015). Precise three-dimensional stereo localization of corner reflectors and persistent scatterers with TerraSAR-X. *IEEE Transactions on Geoscience and Remote Sensing, 53*(4), 1782–1802.

Goel, K., & Adam, N. (2014). A distributed scatterer interferometry approach for precision monitoring of known surface deformation phenomena. *IEEE Transactions on Geoscience and Remote Sensing, 52*(9), 5454–5468.

Goldstein, R. M., Zebker, H. A., & Werner, C. L. (1988). Satellite radar interferometry: Two-dimensional phase unwrapping. *Radio Science, 25*(4), 713–720.

Guarnieri, A. M., & Tebaldini, S. (2008). On the exploitation of target statistics for SAR interferometry applications. *IEEE Transactions on Geoscience and Remote Sensing, 46*(1), 3436–3443.

Gutman, A. S. (1954). Modified Luneberg Lens. *Journal of Applied Physics, 25*(7), 855–859.

Haala, N., & Rothermel, M. (2012). Dense multi-stereo matching for high quality digital elevation models. *Photogrammetrie-Fernerkundung-Geoinformation, 2012*(4), 331–343.

He, M., & He, X. F. (2009). Urban change detection using coherence and intensity characteristics of multi-temporal SAR imagery. 2009 2nd Asian-Pacific Conference on Synthetic Aperture Radar, 840–843.

Hertz, H. (1888). Über die Ausbreitungsgeschwindigkeit der electrodynamischen Wirkungen. *Annalen der Physik, 270*(7), 551–569.

Hetland, E. A., Musé, P., Simons, M., Lin, Y. N., Agram, P. S., & DiCaprio, C. J. (2012). Multiscale InSAR time series (MInTS) analysis of surface deformation. *Journal of Geophysical Research: Solid Earth, 117*(B2).

Hirschmuller, H. (2007). Stereo processing by semiglobal matching and mutual information. *IEEE Transactions on Pattern Analysis and Machine Intelligence, 30*(2), 328–341.

Hooper, A., Zebker, H. A., Segall, P., & Kampes, B. (2004). A new method for measuring deformation on volcanoes and other natural terrains using InSAR persistent scatterers. *Geophysical Research Letters, 31*(23): 1–5.

Hu, X., Wang, T., & Liao, M. (2013). Measuring coseismic displacements with point-like targets offset tracking. *IEEE Geoscience and Remote Sensing Letters, 11*(1), 283–287.

Hülsmeyer, C. (1904). *Verfahren um entfernte metallische Gegenstände mittels elektrischer Wellen einem Beobachter zu melden.*

Jaboyedoff, M., Oppikofer, T., Abellán, A., Derron, M.-H., Loye, A., Metzger, R., & Pedrazzini, A. (2010). Use of LIDAR in landslide investigations: A review. *Natural Hazards, 61*(1), 5–28.

Jao, Jen King. (2001). Theory of synthetic aperture radar imaging of a moving target. *IEEE Transactions on Geoscience and Remote Sensing, 39*(9), 1984–1992.

Jendryke, M., Balz, T., McClure, S. C., & Liao, M. (2017). Putting people in the picture: Combining big location-based social media data and remote sensing imagery for enhanced contextual urban information in Shanghai. *Computers, Environment and Urban Systems, 62*, 99–112.

Jin, M. Y., & Wu, C. (1984). A sar correlation algorithm which accommodates large-range migration. *IEEE Transactions on Geoscience and Remote Sensing, GE-22*(6), 592–597.

Kampes, B. (2006). *Radar Interferometry—Persistent Scatterer Technique.* Dordrecht: Springer.

Kant, I. (1786). *Metaphysische Anfangsgründe der Naturwissenschaft.* J.F. Hartknoch.

Knott, E. F., Shaeffer, J. F., & Tuley, M. T. (2004). *Radar Cross Section.* SciTech Publishing Inc.

Kraus, T., Bräutigam, B., Mittermayer, J., Wollstadt, S., & Grigorov, C. (2016). TerraSAR-X staring spotlight mode optimization and global performance predictions. *IEEE Journal of Selected Topics in Applied Earth Observations and Remote Sensing, 9*(3), 1015–1027.

Krieger, G., Moreira, A., Fiedler, H., Hajnsek, I., Werner, M., Younis, M., & Zink, M. (2007). TanDEM-X: A satellite formation for high-resolution SAR interferometry. *IEEE Transactions on Geoscience and Remote Sensing, 45*(11), 3317–3341.

Krieger, G., Zink, M., Bachmann, M., Brautigam, B., Schulze, D., Martone, M., Rizzoli, P., Steinbrecher, U., Antony, J. M. W., De Zan, F., et al. (2013). TanDEM-X: A radar interferometer with two formation flying satellites. *Acta Astronautica, 89*(2013), 83–98.

Kuo, J. M., & Chen, K. S. (2003). The application of wavelets correlator for ship wake detection in sar images. *IEEE Transactions on Geoscience and Remote Sensing, 41*(6), 1506–1511.

La Prade, G. (1963). An analytical and experimental study of stereo for Radar. *Photogrammetric Engineering, 29*(2), 294–300.

Lanari, R., Mora, O., Manunta, M., Mallorqui, J. J., Berardino, P., & Sansosti, E. (2004). A small-baseline approach for investigating deformations on full-resolution differential sar interferograms. *IEEE Transactions on Geoscience and Remote Sensing, 42*(7), 1377–1386.

Leberl, F., Domik, G., Raggam, J., Cimino, J., & Kobrick, M. (1986). Multiple incidence angle SIR-B experiment over Argentina: Stereo-radargrammetric analysis. *IEEE Transactions on Geoscience and Remote Sensing, GE-24*(4), 482–491.

Lee, J. S., & Pottier, E. (2009). *Polarimetric Radar Imaging: From Basis to Applications.* Boca Raton: CRC Press.

Lee, J. S. (1983). Digital image smoothing and the sigma filter. *Computer Vision, Graphics, and Image Processing, 24*(2), 255–269.

Liao, M., Balz, T., Rocca, F., & Li, D. (2020). Paradigm changes in surface-motion estimation from SAR. *IEEE Geoscience and Remote Sensing Magazine, 8*(1), 8–21.

Lopes, A., Nezry, E., Touzi, R., & Laur, H. (1990). Maximum a posteriori speckle filtering and first order texture models in sar images. *10th Annual International Symposium on Geoscience and Remote Sensing*, 2409–2412.

Luneburg, R. K. (1944). *Mathematical Theory of Optics.* Doctoral dissertation. Brown University. Providence, Brown University.

McConnell, R., Kwok, R., Curlander, J. C., Kober, W., & Pang, S. S. (1991). Correlation and dynamic time warping: Two methods for tracking ice floes in SAR Images. *IEEE Transactions on Geoscience and Remote Sensing, 29*(6), 1004–1012.

Manunta, M., De Luca, C., Zinno, I., Casu, F., Manzo, M., Bonano, M., Fusco, A., Pepe, A., Onorato, G., Berardino, P., De Martino, P., & Lanari, R. (2019). The parallel SBAS approach for Sentinel-1 interferometric wide swath deformation time-series generation: Algorithm description and products quality assessment. *IEEE Transactions on Geoscience and Remote Sensing, 57*(9), 6259–6281.

Massonnet, D., Rossi, M., Carmona, C., Adragna, F., Peltzer, G., Feigl, K., & Rabaute, T. (1993). The displacement field of the Landers earthquake mapped by radar interferometry. *Nature, 364*(6433), 138–142.

Maxwell, J. C. (1865). A dynamical theory of the electromagnetic field. *Philosophical Transactions of the Royal Society, 155*, 459–512.

McConnell, R., Kwok, R., Curlander, J. C., Kober, W., & Pang, S. S. (1991). Correlation and dynamic time warping: Two methods for tracking ice floes in SAR Images. *IEEE Transactions on Geoscience and Remote Sensing, 29*(6), 1004–1012.

Michel, R., Avouac, J. P., & Taboury, J. (1999). Measuring ground displacements from SAR amplitude images: Application to the Landers Earthquake. *Geophysical Research Letters, 26*(7), 875–878.

Milillo, P., Minchew, B., Simons, M., Agram, P., & Riel, B. (2017). Geodetic imaging of time-dependent three-component surface deformation: Application to tidal-timescale ice flow of Rutford Ice Stream, West Antarctica. *IEEE Transactions on Geoscience and Remote Sensing, 55*(10), 5515–5524.

Minh, D. H. T., Hanssen, R., & Rocca, F. (2020). Radar interferometry: 20 years of development in time series techniques and future perspectives. *Remote Sensing, 12*(9), 1364.

Montazeri, S., Gisinger, C., Eineder, M., & Zhu, X. X. (2018). Automatic detection and positioning of ground control points using TerraSAR-X multiaspect acquisitions. *IEEE Transactions on Geoscience and Remote Sensing, 56*(5), 2613–2632.

Mullissa, A. G., Perissin, D., Tolpekin, V. A., & Stein, A. (2018). Polarimetrybased distributed scatterer processing method for psi applications. *IEEE Transactions on Geoscience and Remote Sensing, 56*(6), 3371–3382.

Oersted, H. C. (1820). Experiments on the effects of a current of electricity on the magnetic needle. *Annals of Philosophy, 16*, 273–276.

Oliver, C., & Quegan, S. (2004). *Understanding Synthetic Aperture Radar Images*. Raleigh: SciTech Publishing Inc.

Osmanoglu, B., Sunar, F., Wdowinski, S., & Cabral-Cano, E. (2016). Time series analysis of InSAR data: Methods and trends. *ISPRS Journal of Photogrammetry and Remote Sensing, 115*(100), 90–102.

Paillou, P., & Gelautz, M. (1999). Relief reconstruction from sar stereo pairs: The "optimal gradient" matching method. *IEEE Transactions on Geoscience and Remote Sensing, 37*(4), 2099–2107.

Pathier, E., Fielding, E. J., Wright, T. J., Walker, R., Parsons, B. E., & Hensley, S. (2006). Displacement field and slip distribution of the 2005 Kashmir earthquake from SAR imagery. *Geophysical Research Letters, 33*(20), 269.

Perissin, D., & Ferretti, A. (2007). Urban-target recognition by means of repeated spaceborne SAR images. *IEEE Transactions on Geoscience and Remote Sensing, 45*(12), 4043–4058.

Perissin, D., & Wang, T. (2011). Repeat-pass SAR interferometry with partially coherent targets. *IEEE Transactions on Geoscience and Remote Sensing, 50*(1), 271–280.

Potin, P., Rosich, B., Miranda, N., & Grimont, P. (2018). Sentinel-1A/-1B mission status. EUSAR 2018—12th European Conference on Synthetic Aperture Radar.

Prati, C., & Rocca, F. (1992). Range Resolution Enhancement with Multiple SAR Surveys Combination. *IGARSS '92 International Geoscience and Remote Sensing Symposium*, 1576–1578.

Prats-Iraola, P., Scheiber, R., Marotti, L., Wollstadt, S., & Reigber, A. (2012). TOPS interferometry with TerraSAR-X. *IEEE Transactions on Geoscience and Remote Sensing, 50*(8), 3179–3188.

Preiss, M., & Stacy, N. (2006). *Coherent Change Detection: Theoretical Description and Experimental Results*. Edinburgh: Defence Science and Technology Organisation.

Raggam, H., Gutjahr, K., Perko, R., & Schardt, M. (2010). Assessment of the stereo-radargrammetric mapping potential of TerraSAR-X multibeam spotlight data. *IEEE Transactions on Geoscience and Remote Sensing, 48*(2), 971–977.

Raney, R. K., Runge, H., Bamler, R., Cumming, I. G., & Wong, F. H. (1994). Precision SAR processing using chirp scaling. *IEEE Transactions on Geoscience and Remote Sensing, 32*(4), 786–799.

Rayleigh, F. R. S. (1879). Investigations in optics, with special reference to the spectroscope. *The London, Edinburgh, and Dublin Philosophical Magazine and Journal of Science, 8*(49), 261–274.

Rodriguez-Cassola, M., Prats-Iraola, P., De Zan, F., Scheiber, R., Reigber, A., Geudtner, D., & Moreira, A. (2015). Doppler-related distortions in TOPS SAR images. *IEEE Transactions on Geoscience and Remote Sensing, 53*(1), 25–35.

Rosenfield, G. (1968). Stereo radar techniques. *Photogrammetric Engineering, 34*(6), 586–594.

Ruch, J., Wang, T., Xu, W., Hensch, M., & Jonsson, S. (2016). Oblique rift opening revealed by reoccurring magma injection in central Iceland. *Nature Communications, 7*(1), 12352.

Scambos, T. A., Dutkiewicz, M. J., Wilson, J. C., & Bindschadler, R. A. (1992). Application of image cross-correlation to the measurement of glacier velocity using satellite image data. *Remote Sensing of Environment, 42*(3), 177–186.

Scheiber, R., & Moreira, A. (2000). Coregistration of interferometric SAR images using spectral diversity. *IEEE Transactions on Geoscience and Remote Sensing, 38*(5), 2179–2191.

Serafino, F. (2006). SAR image coregistration based on isolated point scatterers. *IEEE Geoscience and Remote Sensing Letters, 3*(3), 354–358.

Siddique, M. A., Wegmüller, U., Hajnsek, I., & Frey, O. (2016). Single-look sar tomography as an add-on to psi for improved deformation analysis in urban areas. *IEEE Transactions on Geoscience and Remote Sensing, 54*(10), 6119–6137.

Singleton, A., Li, Z., Hoey, T., & Muller, J. P. (2014). Evaluating sub-pixel offset techniques as an alternative to D-InSAR for monitoring episodic landslide movements in vegetated terrain. *Remote Sensing of Environment, 147*, 133–144.

Stephens, M. A. (1970). Use of the Kolmogorov–Smirnov, Cramér–von Mises and related statistics without extensive tables. *Journal of the Royal Statistical Society: Series B (Methodological), 32*(1), 115–122.

Teunissen, P. J. G. (1995). The least-squares ambiguity decorrelation adjustment: A method for fast GPS integer ambiguity estimation. *Journal of Geodesy, 70*(1), 65–82.

Toutin, T., & Chenier, R. (2009). 3-D radargrammetric modeling of RADARSAT-2 ultrafine mode: Preliminary results of the geometric calibration. *IEEE Geoscience and Remote Sensing Letters, 6*(2), 282–286.

Varian, R. H., & Varian, S. F. (1939). A high frequency oscillator and amplifier. *Journal of Applied Physics, 10*(5), 321–327.

Viola, P., & Wells, W. M. (1997). Alignment by maximization of mutual information. *International Journal of Computer Vision, 24*(2), 137–154.

von Schelling, F. W. J. (1797). *Ideen zu einer Philosophie der Natur.* Leipzig: Breitkopf und Härtl.

Wang, J., Balz, T., & Liao, M. (2016). Absolute geolocation accuracy of high-resolution spotlight TerraSAR-X imagery —validation in Wuhan. *Geo-spatial Information Science, 19*(4), 1–6.

Wang, R. (2013). 3D building modeling using images and LiDAR: A review. *International Journal of Image and Data Fusion, 4*(4), 273–292.

Washaya, P., Balz, T., & Mohamadi, B. (2018). Coherence change-detection with Sentinel-1 for natural and anthropogenic disaster monitoring in urban areas. *Remote Sensing, 10*(7): doi:10.3390/rs10071026.

Weihing, D., Hinz, S., Meyer, F., Laika, A., & Bamler, R. (2006). Detection of along-track ground moving targets in high resolution spaceborne SAR images. *Proceedings of ISPRS,* 81–86.

Werner, C., Wegmuller, U., Strozzi, T., & Wiesmann, A. (2003). Interferometric point target analysis for deformation mapping. *Proceedings of 2003 IEEE International Geoscience and Remote Sensing Symposium. 7,* 4362–4364.

Wiley, C. (1951). Pulsed Doppler Radar Methods and Means. United States Patent and Trademark office.

Xie, H., Pierce, L. E., & Ulaby, F. T. (2001). Mutual Information Based Registration of SAR Images. 1995 IEEE International Geoscience and Remote Sensing Symposium, 4028–4031.

Yague-Martinez, N., Prats-Iraola, P., Rodriguez Gonzalez, F., Brcic, R., Shau, R., Geudtner, D., Eineder, M., & Bamler, R. (2016). Interferometric processing of Sentinel-1 TOPS data. *IEEE Transactions on Geoscience and Remote Sensing, 54*(4), 2220–2234.

Young, T. (1802). On the theory of light and colours. *Philosophical Transactions of the Royal Society of London, 92,* 12–48.

Zebker, H. A., & Villasenor, J. (1992). Decorrelation in interferometric radar echoes. *IEEE Transactions on Geoscience and Remote Sensing, 30*(5), 950–959.

Zhang, L., Lu, Z., Ding, X., Jung, H.-S., Feng, G., & Lee, C.-W. (2012). Mapping ground surface deformation using temporarily coherent point SAR interferometry: Application to Los Angeles Basin. *Remote Sensing of Environment*, *117*(100), 429–439.

Zhu, X. X., Baier, G., Lachaise, M., Shi, Y., Adam, F., & Bamler, R. (2018). Potential and limits of non-local means InSAR filtering for TanDEM-X high-resolution DEM generation. *Remote Sensing of Environment*, *218*, 148–161.

Index

www.ingramcontent.com/pod-product-compliance
Lightning Source LLC
Chambersburg PA
CBHW050601190326
41458CB00007B/2128